Even if it's the result of one year's work, this book is not free of inaccuracies. Feel free to contact me about any issue with the text.

7

Illustration Index

13

1 Introduction to 802.11ac

1.1 Introduction
This book has two goals, first to summarize and present the huge 802.11ac amendment in a meaningful context. That was achieved in the first version of this book. This new version adds explanations about Linux implementation of 802.11ac in open source drivers.

1.1.1 Brief overview of 802.11 and 802.11ac

802.11 is a system for data transmission by radio, between electronic equipment in an area with a radius of a few tens of meters. It's was originally designed as a wireless LAN at a time where cabled networks were offering throughput of nearly 50 Mbits/sec with an excellent privacy. 802.11 was and is still seen as a very low cost alternative to cabled networks, however technologies profoundly evolved during the last 20 years and so did 802.11.

The major features of all generations of 802.11 are:

- It enables LAN creation at low cost in home or small offices.

- It's a miniaturized technology that is almost transparent to the user.

- While still not perfect, it's relatively easy to configure and use by non experts. It has virtually no maintenance.

- It is relatively privacy friendly for home or small office even if it is far from perfect in this area.

- It works on many types of hardware and operating systems. Thanks to Wi-Fi Alliance products from different manufacturers are interworking.

IEEE 802.11ac is a recent standard that offers high throughput, at least 500 Megabits/sec and up to several Gigabits/sec depending on hardware in the 5 GHz band. This specification will allow much higher speeds than the previous standard. These results were obtained using a wider radio channel, up to 8 streams MIMO, and a highly sophisticated modulation (256 QAM). Actually 802.11ac allows throughput ten times higher than 802.11n and it really delivers it because it relies on simpler technologies that really work. However 256 QAM requires a radio environment which will be probably difficult to find in practice. The design of MIMO is much simpler than in 802.11n. In addition not only performance is considered in the standard but resilience to interference or noise is improved. There are coexistence mechanisms for the channels with a width of 20/40/80, and 160 MHz between the stations 802.11ac and 802.11a/n stations. The spectrum is also better used thanks to MU-MIMO.

This standard is somewhat a reaction to manufacturers' frustrations during the design of IEEE 802.11n where a huge amount of ideas had been brought in by academics, few having been rejected and lot of heterogeneous proposals adopted. Chipset designers considered that at the end of the standardization work, 802.11n was a too complex standard as it was at the same time, very innovative but still interoperable with the legacy standards 802.11g (2.4GHz) and 802.11a (5GHz). In addition some of their clients complained that 802.11n instead of offering a throughput of hundreds of megabits as advertized, offered only an incremental throughput improvement over the previous generation.

This complexity of 802.11n is evident by simply considering in one hand the number of MCS (77 including 32 which are considered as realistic and only 8 which are mandatory for a station) and in the other hand the 8 MCS of 802.11ac, with six of them which exist already in 802.11n.

So in May 2007 as IEEE 802.11n chipsets became available on the market, the IEEE 802.11 working group set up a new study group called "very high speed" or "VHT", for the improvement of the IEEE 802.11 throughput. Very soon two proposals emerged because of different preferences about the operating radio band. One of them concerned the use of the 5 GHz band. In fact an 802.11n 40 MHz channel occupies almost all the 2.4 Megahertz bandwidth so anyway the 2.4GHz is nearly useless for high throughput and 802.11ac designers wanted a canal with an 80 Megahertz width to be mandatory. The other was more interested in the 60 GHz band. The reason here is that the millimeter wave (30 GHz to 300 GHz) band allows naturally much higher rates than the centimeter band.

The VHT study group initiated the birth of two IEEE 802.11 working group, one for the 5 GHz band (802.11ac) and one for the 60 GHz band (802.11ad).

The scope of 802.11ac includes:

- Support of a minimum throughput of at least 500 megabits per second simply with larger and more efficient modulation on radio channels.
  - Larger channels width up to 80 MHz to 160 MHz against "only" 40 MHz width at maximum in 802.11n. 80 MHz width channels support is mandatory; 160 MHz channels support is optional. Because of the difficulty to find two channels at 80 MHz that are contiguous (only 8 20MHz channels are not subject to DFS), a 160 MHz channel can be composed of two non-contiguous channels at 80 MHz.

  - Support for a new modulation: 256 QAM, with two coding rates 3/4 and 5/6, these modes being optional (vs. 64 QAM, rate 5/6 in 802.11n at its best)

- Performance boost with MIMO up to at least 1 Gbps with a maximum of 6.9 Gbps. 802.11ac uses more MIMO streams: Up to 8. Naturally this assumes there are as much available antennas. It also uses Multi-User MIMO (MU-MIMO). It is a new way to create several geographical sections within a cell thanks to "Space Division Multiple Access" (SDMA). With MU-MIMO several stations can now transmit simultaneously with their access point. It is based

on MIMO, but it mandates the use of several sets of antennas.
- Backward compatibility and coexistence with 802.11a/n devices in the 5 GHz band, which meant encapsulating VHT preamble in an 802.11a frame.

As was the case with the 802.11n products, 802.11ac products have emerged on the market well before the standard was officially adopted.
As early as the mid-2012 one could buy routers or USB sticks from brands such as NETGEAR because chipsets were available since February 2012 at the usual manufacturers (Broadcom, Qualcomm, Intel and Mediatek). In most implementations an 802.11n chip accompanies the 802.11ac chip for operation in 2.4GHz band. However, these chipsets have only the 802.11ac mandatory features, and it is possible that optional, richer features may be included over the years. The Wi-Fi Alliance introduced tests and recommendations for interoperability during two "waves" during the year 2013, one around April, and the other in mid-year.

1.1.2 OVERVIEW OF THE IEEE AND 802.XX STANDARDS

The IEEE is a non-commercial organization, which any engineer can join freely provided that she/he is working in the sector of electricity, electronics or informatics.

The IEEE reflects within ad hoc groups, on themes relevant to the profession where the decision-making is collegial. Its goal is to advance humanity welfare by inventing and sharing new technologies.

The IEEE communicates through journals such as IEEE Spectrum, conferences and standards.

1.1.3 802.11 AND ITS NORMATIVE ENVIRONMENT.

The 802.11 standard describes the interfaces and operation of hardware and software resources that provide the ability to exchange data using a radio channel.

Although the term 802.11 is often used interchangeably with the term Wi-Fi, 'Wi-Fi' is a term used to indicate compliance with interoperability specifications that written by a group of manufacturers and users: the Wi-Fi Alliance.

The 802.11 specifications do not guarantee interoperability between different suppliers, because as in all specifications, some functions are optional and implementations are not described, although the algorithms and the use cases or impact studies are often part of the standard design. Probably this separation helps to deal with one problem at a time, first there is a work on specification design, and after that interoperability between manufacturers is checked in a second phase. The 802.11 specifications are developed over several years, but almost every year sees the arrival of a new specification or an amendment to an existing standard.

There are also several troubling aspects about IEEE specifications. For example it is not clear what is mandatory or not apart the main features. For example the Linux mac80211 layer implements only a few part of the 802.11 MAC, is it really compliant with it? Another thing is that only a few parts of the specification use a formal description and anyway nobody verifies the compliance of an 802.11 device to the specification. Again in Linux, there are several 802.11ac drivers that each implements different sets of features, but always a subset of the specification. The references to the specification are hardly explicit in the source code, it may be because of companies' reluctance to openness but it may also point to some inherent flaws in IEEE 802.11 standardization methods as some have already pointed out.

The specification also incorporates hundreds of patents with most not disclosed explicitly at IEEE. So sometimes implementers prefer to consider this document only as a guide. The Wi-Fi Alliance creation may have been initiated by such considerations because it prefers to focus on interoperability which is certainly a central concern in communication equipment.

While the previous revision (2007) of 802.11 standard incorporated the OFDM based 802.11g and 802.11a amendments in the standard, the current revision of 802.11: IEEE Std 802.11-2012, incorporates the following changes to the 2007 revision:

- 802.11 k: Radio resource measurement in wireless LANs

- 802.11r: Fast BSS Transition (roaming from one BSS to the other)

- 802.11y 3650 MHz operation in the U.S.

- 802.11w: protected management frames

- 802.11n: enhancements for higher throughput

- 802: access in vehicular wireless environments

- 802.11z: extensions of direct link Setup (DLS)

- 802.11v: Wireless Network Management

- 802.11u: Interworking with external networks

- 802.11s: Mesh Networking

1.1.4 WI-FI ALLIANCE

The Wi-Fi Alliance is an association of manufacturers which aims at promoting the use of the 802.11 standard on several markets. It was created because of the increasing number of interoperability issues between materials of different origins as the market was growing. It was recognized that obtaining a consensus about those issues, within the IEEE 802.11 was difficult and that was limiting the acceptability by users.

Wi-Fi recommendations generally apply only to a subset of the 802.11 specifications. It aims at expanding the wireless products market by making the promotion of some user oriented features such as interoperability between different providers, but sometimes entirely new features are created to tackle urgent problems such as security. This focus on the market is different from IEEE focus on innovation. It could be said that IEEE is the place where academic's proof of concepts are transformed in industry grade products, whereas Wi-Fi Alliance must meet the market demands and therefore have interest in producing devices in the short term, in a cost-effective manner.

Academics have interest to introduce advanced features and concepts, and the cost of implementation is often not their primary concern. Some IEEE members, including academics have interest to file patents on standard technologies.

The purpose of the Wi-Fi Alliance is explicitly to promote interoperability without creating new technologies. There is therefore less problem of common understanding between members with the same goals.

The Wi-Fi Alliance operates different techniques to achieve this goal.

- It promotes a label indicating compatibility with a subset of 802.11. Compatibility is tested with respect to reference devices.

- It creates and promotes certain features such as WPA or PASSPOINT.

It is worth noting that the names known to the public (WEP/WPA/etc..) are most often coming from Wi-Fi Alliance, the technical names proposed by the IEEE do not reach a wide audience.

Here is what Wi-Fi Alliance proposes as features for 802.11ac based chips:

The **mandatory** features in Phase 1 are:

- 5GHz operation (2.4 GHz is excluded)

- 20, 40, and 80MHz channels

- 1 spatial stream but 2 spatial streams for non-mobile devices, because less channel state information dialogue in fixed stations.

- MCS 0-7 (BPSK, r=1/2 through 64-QAM, r=5/6)

- VHT A-MPDU delimiter for RX and TX for single MPDU

- A-MPDU in reception

- A-MPDU in transmission

- Clear Channel Assessment (CCA) on Secondary channel

- CTS with bandwidth signaling in response to RTS with bandwidth signaling

The **optional** features in Phase 1:

- 2 or 3 spatial streams client
- 2 spatial streams AP for Mobile Access Points
- 3 spatial stream for fixed Access Points
- MCS 8 (256-QAM, r=3/4)
- MCS 8-9 (256-QAM, r=3/4 and r=5/6)
- Short Guard Interval (GI)
- AP STBC in transmission 2x1 (improving SNR by using multiple paths)
- STA STBC in reception 2x1
- Receiving A-MPDU of A-MSDU
- RTS with bandwidth signaling signaling
- Transmit beam-forming (TxBF)
- Low Density Parity Check (LDPC) coding

Phase 2 features:

- 160MHz channels
- 80 + 80MHz channels
- 4 spatial stream for client and AP
- MU-MIMO
- TXOP sharing

- VHT TXOP power save

31

1.1.5 THE 802.11 SOFTWARE STACK

802.11 Standards are part of the IEEE standards, for example the 802.11 LLC layer is derived from the 802.2 layer.

When a user starts the browser on her computer, this browser that belongs to the user land applications, uses a protocol stack that is most often delivered with the operating system. By making calls to this protocol stack, the browser is able to exchange information with a remote computer. A stack example consists of the TCP/IP protocol that allows exchange of information on networks of different nature by fragmenting information into packets, which are sent by a messaging system independent of the networks to which the terminals are connected. For example on an IEEE LAN, the IP layer uses Ethernet services.

Where the computer communicates in 802.11, this stack uses the services of a device driver that will convert these data in radio signals. The device driver sends command and data to the device, where there is usually a LLC/MAC and PHY layers. While receiving information from the media radio the information flow goes from the device, to the device driver, to the TCP/IP stack and from there, to the application.

There are three layers in 802.11:

- LLC (logical link control) is designed to allow multiple applications or multiple networks host, access to resources 802. This layer allows handling different concurrent requests to access a physical asset.

- MAC (media access control) allows several terminals to communicate through a single media. There is therefore an addressing system internal to the MAC layer enabling each terminal to send messages to the address of another terminal. The MAC layer manages the contention to access.

- PHY (physical layer) provides the transformation of the digital information in radio signals, and the other way round.

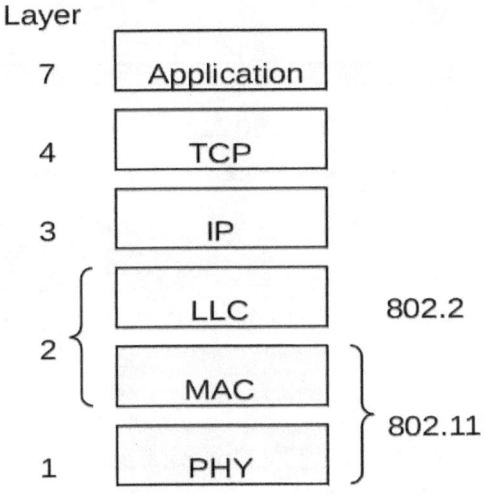

Figure 1 Protocol stack

In Linux this communication between layers is done by using different APIs that evolved over the time. While 10 years ago a precursor was named "Wireless extensions" or "wext", nowadays the communication between the IP stack and the MAC layer goes through the "nl80211" layer like in the figure below:

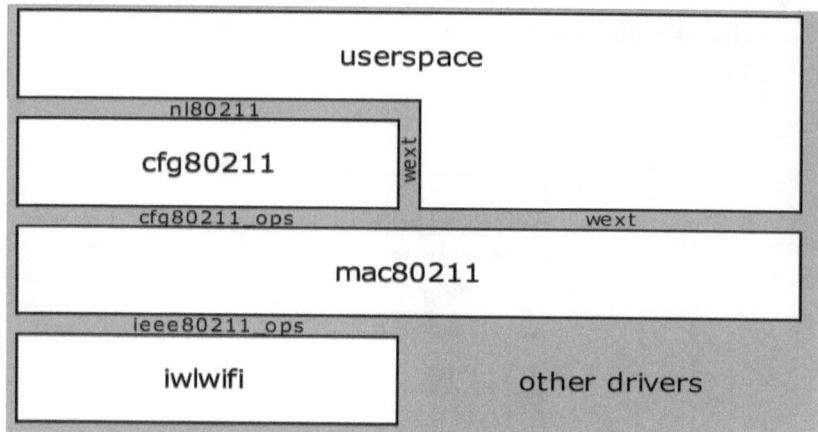

Figure 2: Linux 802.11 stack layering

Mac80211 is a Linux kernel module which offers support of a part of 802.11 MAC to 802.11 chipsets based on SoftMAC. It has a callback mechanism which permits a specific chipset to use its own procedures, instead of those of mac80211 if it is preferable. In the figure above the callbacks are called cfg80211_ops or mac80211_ops.

mac80211 is a framework which driver developers can use to write drivers for SoftMAC wireless devices. It originates in FreeBSD net802.11.

SoftMAC devices allow applications to have a finer control of the hardware, eventually allowing 802.11 frame management to be done in software. Most 802.11 devices today tend to be of this type: FullMAC is used only in low end or embedded devices.

mac80211 implements the cfg80211 callbacks for SoftMAC devices, mac80211 then depends on cfg80211 for both registration to the networking subsystem and for configuration. Configuration is handled by cfg80211 both through nl80211. "wireless extensions" is an obsolete library similar in goal to nl80211. Operations are done through "ops" codes; each code implements a different operation such as "associate".

In mac80211 the MLME is done in the kernel for station mode (STA) and in userspace for AP mode (hostapd).

802.11 Standards define a "frame" type for the transmission of data as well as the management and control of the radio hardware. As we will see in details later, the frames are divided into sections. Each frame roughly consists of a header, the payload and frame check sequence (FCS). Some frames may not contain useful data but signalization such as RTS/CTS/ACK. The three layers each manage different issues and therefore each adds information to the package from the top layer.

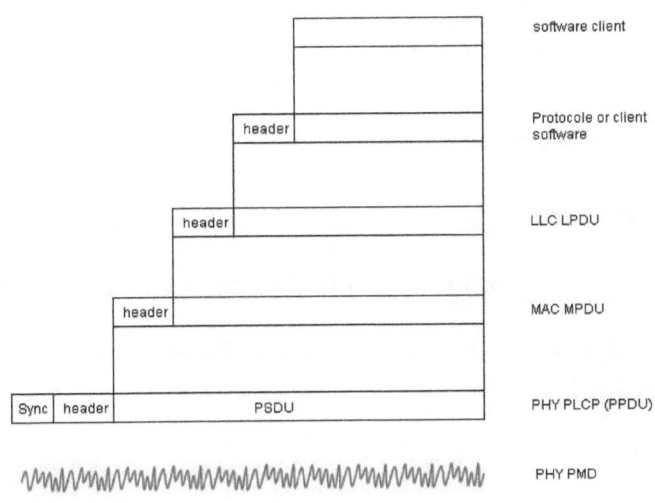

Figure 5 Protocol layers encapsulation

The IEEE 802.11 standards thus live into those three layers:

1) LLC data link layer:

- 802.11e: QoS

- 802.11f: inter-Access point protocol roaming

- 802.11i: security

2) The media access layer:

- 802.3

3) Physical layer:

- Infra-red: currently only for the obsolete standard 802.11

- FHSS: Only for the obsolete standard 802.11

- DSSS: For the obsolete standard 802.11b

- OFDM: Standards 802.11a, and 802.11n and 802.11ac

The following frequency bands are used in the various amendments:

- 2.4 GHz (IEEE 802 .11b, 802. 11g and 802.11n)

- 5 GHz (IEEE 802.11a, 802.11n and 802.11ac)

- Very high throughput WLAN 60 GHz (IEEE 802.11ad)

- White spaces (.11af, different bands, especially around 800 MHz)

- Sub GHz (802.11ah)

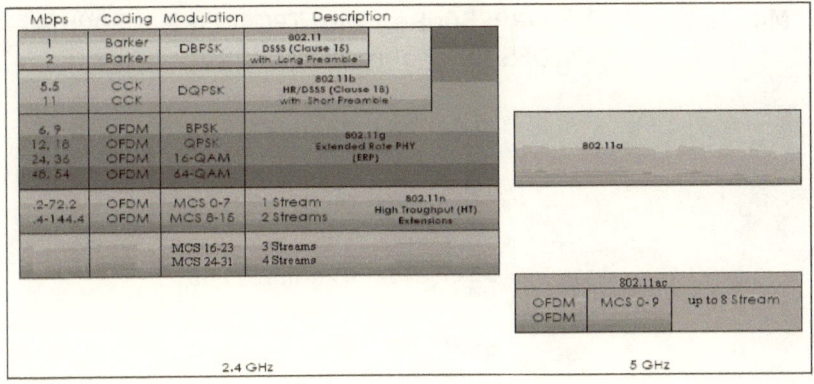

Figure 6 IEEE 802.11 standards

802.11ac operates only in 5GHz band and in the following channels:

/* UNII 1 */	EIRP: 23dBm, indoor only
	5180 MHz Channel 36
	5200 MHz Channel 40
	5220 MHz Channel 44
	5240 MHz Channel 48
/* UNII 2 */	EIRP: 30dBm, with TPC if EIRP > 500mw
	5260 MHz Channel 52
	5280 MHz Channel 56
	5300 MHz Channel 60
	5320 MHz Channel 64

/* "Middle band" */	EIRP: 30dBm, with TPC if EIRP > 500mw, DFS for channels 120 to 128.
	5500 MHz Channel 100
	5520 MHz Channel 104
	5540 MHz Channel 108
	5560 MHz Channel 112
	5580 MHz Channel 116
	5600 MHz Channel 120
	5620 MHz Channel 124
	5640 MHz Channel 128
	5660 MHz Channel 132
	5680 MHz Channel 136
	5700 MHz Channel 140
/* UNII 3 */	EIRP: 36dBm (except channel 165 Tx power < 30dBm)
	5745 MHz Channel 149
	5765 MHz Channel 153
	5785 MHz Channel 157
	5805 MHz Channel 161
	5825 MHz Channel 165

1.1.6 ABOUT THIS BOOK.

In this book we review the two layers MAC, PHY that are the subject of 802.11ac amendment. These two layers are going to be studied in detail based on the draft 6 of 802.11ac; it is probably one of last drafts before the IEEE administrative process leading to the official publication in late 2013 and it is also the draft which will serve as support to the Wi-Fi Alliance for the second wave of certification. This book gives a quite accurate description of the standard but this is absolutely not a lavish copy because it also tries to introduce the underlying concepts, give some context, and takes into account the constraints which are those of the real life.

A difficulty in this book sits in this layered structure because some subjects are naturally difficult to reduce to a layer, for example CCA seems a pure MAC topic about sharing the radio media access issue, and not a pure PHY problem but this is more complicated with topics such as the aggregation of channels, which one would think is a pure PHY topic but it implies a particular form of acknowledge, so it belongs also to the MAC level. The index at the end of the book should solve this problem. This book still uses the PMD notion even if 802.11ac designers especially at Intel made it disappear from the amendment. Other amendments still use this PMD notion. This is an indication that layering is often artificial, even if it helps structure learning or design.

1.2 802.11AC TECHNOLOGIES

1.2.1 LARGER CHANNELS
The increase in throughput occurs primary via using larger radio channels. The width of channels was increased to 80 MHz and eventually to 160 MHz against 40 MHz maximum in 802.11n and 20MHz practically in 2.4GHz. Support for 80 MHz sized channels is mandatory in 802.11ac; 160 MHz channel support is optional. A 160 MHz channel can even be composed of two non-contiguous channels at 80 MHz. This is because of the difficulty in finding two contiguous 80 MHz channels. Actually there will be only three or four channels really without interference in 802.11ac depending where it is used and not all channels are equals in permitted power or existence of DFS and thus the 5GHz band will saturate as fast as the 2.4 GHz band.

1.2.2 MIMO, SPATIAL DIVERSITY, BEAM-FORMING AND MU-MIMO
An important design goal of 802.11ac was to improve the performance in MIMO with multiple users. MIMO is a mode where multiple electromagnetic signals are received at the same time on the same frequency band. MIMO is an umbrella term that encompass very different technologies and goals:

* Multiple spatial streams increases throughput.

* Spatial diversity increases SNR (signal/noise ratio).

* Beam-forming increases SNR only in specific places.

* MU-MIMO enables several simultaneous communication is different directions.

Let's get some basics facts first. Multiple propagation paths enable the possibility to additively combine energy received by each antenna, MIMO will thus increase the signal to noise ratio and therefore the throughput. To work well each emitter antenna must be driven by a separate amplifier and each signal must be coded in a different way from its homologue to enable an easy separation at the receiver. So the limiting factor is the number of antennas on the emitter but a station is in turn sometimes an emitter and sometimes a receiver.

On the receiver side, it's possible to add the signal coming from several antennas to obtain a better SNR (signal to noise ratio). This is an improvement on pre-MIMO chipsets were several antennas were already used but only the antenna receiving the best signal was used.

1.2.3 BEAM-FORMING

Beam-forming has been introduced with 802.11n but there were few implementations because, first this feature was optional and second it required cooperation at both the transmitter and the receiver. This function is also optional in 802.11ac but only the transmitter "forms" the beam. The receiver has "just" to inform the transmitter about the characteristics on the channel. 802.11ad still uses a more complex beam forming procedure.

Beam-forming is very similar to a usual user experience with AM/FM radio when one search the best physical place and antenna orientation to receive a station, because probably in both cases interferences, between several radio reflections on the environment, are involved.

Beam-forming transmissions are adjusted to the characteristics of the path between the access point and each station. This involves a reporting protocol called "sounding protocol" (also optional) support. The emitter sends a special packet with no data and the receiver sends back what radio media characteristics can be inferred about the path. IEEE still uses the term "channel sounding" but actually it's the individual paths characteristics between two antennas that are reported.

Of course it is unlikely to see eight antennas on a station when you know that most mobile phones have only a single 802.11 antenna and an access point only three or four antennas. "Pre-draft" chipsets have only from one to three antennas and it is whispered that the Wi-Fi Alliance will not go beyond four antennas in its certification.

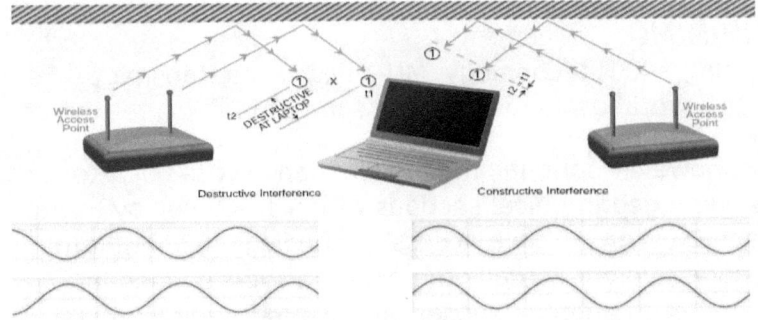

Figure 7 Interferences

1.2.4 MU-MIMO

802.11 being a LAN technology, MU-MIMO helps transmit a content to several users of the same sub-group.

This new, however optional, mode allows an access point to create several geographical sections within a BSS cell by using the "Space Division Multiple Access" technology (SDMA). It may sound impressive but actually the antenna set is divided in subsets and each antenna subset targets a different user with beam-forming. Inside a given beam, MIMO is used. There is no need to have several directional antennas for those different flows, all is needed is several omnidirectional antennas. This combination of interference management technologies and phase patterns between antennas should provide a better result than simply using only one of those technologies.

By changing slightly the phase of the signals at the emitter antennas, some stations that were previously in blind spots can now receive the signal quite clearly; conversely the stations that were previously in a good spot now experience a much worse link quality. We can now address different groups separately as if they were not in the same cell. This helps to improve the global throughput as previously all stations had to share the bandwidth and now only the stations in the same group have to share it.

Common sense applies as the number of simultaneous communications multiplied by the number of paths at one end can't be higher than the number of antennas. In real life the benefits are much lower than we could expect from the mathematical formulation which is about a perfect situation. The angular section or "lobe" is very imperfect, the lobe width is approximately 45° and this lobe does not exist at short distances (uniform propagation) and its effect is fairly small at great distance.

It should be noted that in 802.11ac only the access point is at the origin of the MU-MIMO exchange, it is called DL-MU-MIMO (Downlink multi-user MIMO). One practical reason is that in

MIMO it takes at least several antennas to achieve simultaneous exchanges (spatial stream) and MU-MIMO is worsening this constraint. So to exercise MU-MIMO toward four destinations at the same time, there must be at least eight antennas which is a lot even for an access point.

1.2.5 MCS (MODULATION AND CODING SCHEMES)

The throughput depends not only on the radio channel width and degree of parallelism, but it depends also on the ability to encode more information on the radio carrier in a given bandwidth, thanks to various modulation and coding schemes (MCS).

A MCS is simply a set of parameters determining throughput.

Several MCS are needed because a station may experience various radio conditions so the optimal throughput is not always the most aggressive. 802.11ac supports seven of the 802.11n mandatory MCS modes of (which had a total of 77 MCS!) and it adds two others based on 256-QAM modulation, of 3/4 and 5/6 coding rates (vs. 64-QAM in 802.11n at 5/6 maximum rate), these modes are optional. If this 256-QAM modulation appears to be a technology breakthrough, it only increases the flow of approximately 40%.

Other radio technologies may have better modulations, for example DVB-C2 (DVB for cable) allows using 4096-QAM modulation, a 50% gain over 256 QAM, and future DVB-C2 extensions will even use 16384-QAM and 65536-AQAM modulation. But 802.11 is a low cost technology by design, so it can't use expensive hardware. Optional support for 8 and 9 MCS, suggests that in reality the flow will rarely exceed 325 Mbits/sec per physical stream, and two or three streams will be the norm, leading to throughput slightly below 1GHz in real consumer products.

Modulation type	Number of bits per OFDM symbol
16 QAM	4
64 QAM	6
256 QAM	8

Figure 8 Modulation types

In addition there is a need for error corrections, as raw PHY frames always carry a few errors. Forward Error Correcting Codes help to remove the detrimental effects of errors. It helps improve the SNR at least by a factor two (3db), however the main cause of errors may often lie in the receiver, not in the radio path.

There are two kind of error correcting codes, iterative or not. Iterative codes use additional information to correct errors, as this is not a simple task, it should be done in several rounds where some hypothesis is tested until a satisfying solution is found.
As it uses external information, the iterative codes need less overhead than non-iterative codes. This external information is provided by the Fast Fourier Transform which extract information from the OFDM sub-carriers. 802.11ac uses LDPC as an optional iterative code; otherwise it uses BCC which is a non-iterative forward error correcting code.

MCS	Modu-lation	Coding rate	Through-put	Coded bits per sub-carrier	Coded bits per OFDM symbol	Data bits per OFDM symbol	Sensi-tivity (80 MHz)
0	BPSK	0.50	32.5	1	234	117	−76dBm
1	QPSK	0.50	65	2	468	234	−73dBm
2	QPSK	0.75	97.5	2	468	351	−71dBm
3	16−QAM	0.50	130	4	936	468	−68dBm
4	16−QAM	0.75	195	4	936	702	−65dBm
5	64−QAM	0.67	260	6	1404	936	−60dBm
6	64−QAM	0.75	292.5	6	1404	1053	−59dBm
7	64−QAM	0.83	325	6	1404	1170	−57dBm
8	256−QAM	0.75	390	8	1872	1404	−54dBm
9	256−QAM	0.83	433	8	1872	1560	−52dBm

1.2.6 OTHER FEATURES

As in previous amendments, there are coexistence mechanisms for the large 80/160 MHz channels in 802.11ac stations to coexist with the channels of earlier 5GHz technologies such as in 802.11 a/n stations.

In order to support the newer use cases such as the large data packets transmitted by Web servers or video servers, there is a mandatory support of A-MPDUs in a VHT PPDU. This reflects changes in usage and technologies as in the past, it was usual to have small packets on LANs because hardware were not able to cope with high throughput and memory was costly so there were only small buffers in routers and other network equipment.

1.3 MAIN CONCEPTUAL ELEMENTS OF 802.11AC SIGNAL

In order to qualify as an 802.11ac signal, one must choose between a number of parameters such as:

* The size of channel (20, 40, 80, 160, and 80 + 80), the number of antennas, and the frame configuration. A frame can have several physical modes (mixed mode, legacy, Greenfield...). The channels can be arranged in different ways: (HT 20, HT 40 HT duplicate HT Upper HT lower, VHT 20, VHT 40, VHT 80, VHT 80 + 80, VHT 160).

* There may be multiple streams (paths between an antenna in the emitter and the receiver) which would vastly increase throughput, or spatial diversity could be chosen instead because the SNR is too low, sometimes beam-forming allows better SNR than spatial diversity, the AP may decide to use multiple antennas for MU-MIMO instead of raw performance with one station at a time. Those choices are not specified in 802.11ac, they may have a considerable influence on real life performance, and given that Wi-Fi experience a great diversity in hardware and software the worst case is to be expected in an un-managed situation.

* There may be LDPC or not.

One choice must be made between several MCS, i.e. the combination of parameters to the PHY level that determines the throughput, but this is not usually done by the end user, it is done by rate adaptation algorithms. Some are well known such as Minstrel in Linux, some are proprietary. It should be noted that Minstrel gives excellent results at MAC level, but it ignore opportunities offered by MIMO.

1.3.1 EXAMPLES OF THROUGHPUT IN 802.11AC

Scenario	PHY Link Rate
AP with two antennas, a station with two antennas, channel at 40 MHz, 64QAM	300 Mbit/s (2 * 150Mbits/sec) it is the same capabilities as an 802.11n client and short GI.
AP with an antenna, a station with an antenna, 80 MHz channel, 256 QAM	433 Mbits/s
AP with two antennas, a station with two antennas, 80 MHz, 256 QAM	867 Mbits/s (2 * 433 Mbits/sec)
AP with an antenna, a station with an antenna, 160 MHz, 256 QAM	867 Mbits/s, it uses all 8 non DFS channels!
AP with two antennas, a station with two antennas, 160 MHz, 256 QAM	1.73 Gbit/s (2* 867 Mbits/sec)
AP with four antennas, four stations with an antenna each 80 MHz channel (MU-MIMO)	433 Mbit/s for each station (four streams are used at each receiver to increase the SNR, so to increase the throughput)

AP with eight antennas, 160 MHz channel, and a station with four antennas	3.47 GB/s because only four physical streams are allowed by the four antennas on receiving stations. However if the AP was smart (which it is probably not the case) it could combine four physical streams with beam-forming couples of antennas in such a scheme where each spatial stream takes advantage of two beam forming antennas from the set of 8 antennas. It would require a very sophisticated AP and STA.
AP with eight antennas, 160 MHz channel, and a station with two antennas	1.73 GB/s for a station with two antennas as there are only two physical streams available.
MU-MIMO AP with eight antennas, 160 MHz channel, and two stations with an antenna each	867 Mbit/s for each station with an antenna. However MU-MIMO implies two set of 4 antennas participate in beam-forming, so the SNR could be quite good. But this configuration seems little capable of delivering the theoretical performance in practice.
AP with eight antennas, four stations with two antennas, 160 MHz channel	1.73 GB/s for each station, in theory simultaneously, but each MU-MIMO path has only two antennas involved in it, so the SNR is low.

Figure 9 Some 802.11ac configurations

1.4 802.11 Architecture

1.4.1 802.11 Architecture components
1.4.1.1 The station

The station (often abbreviated in STA) is the most common component of 802.11 networks. A station is a device that provides the functionality of the 802.11 Protocol:

- LLC

- Media access control (MAC).

- Physical layer (PHY),

- A radio media connection.

In General, 802.11 functions are implemented in the hardware and software of an interface card (NIC) card. A station may be a laptop, a phone, a tablet, or an access point (AP). All stations offer 802.11 services such as authentication, privacy and data transport.
Under Linux a NIC always operates in one of the following operating modes (STA, AP, MESH). The mode sets the main functionality of the wireless link. It is possible to run in two modes at the same time.
Any wireless driver is capable of running in the station mode. In BSS mode (infrastructure mode), two STA cannot connect directly to one another. They require a third NIC in AP mode to manage the wireless network. A STA associates to an AP by sending certain management frames to it. This process is called the authentication and association. After the AP sent the successful association- reply, the STA is part of the BSS.

This mode is managed with the aging "wireless extension " WEXT tools "iwconfig" or the newer "iw" tool.

1.4.1.2 Basic Service Set (BSS)

Basic Service Set (BSS) is the basic element of an 802.11 LAN. The BSS consists of a number of stations and an access point. The access point manages all information exchange as well as non-functional aspects such as security.

1.4.1.3 Access Point (AP)

The access point is a station that has a particular role in infrastructure mode. All exchanges are between a station and the access point except in certain modes of the 802.11ac where two stations can directly exchange content. Otherwise there is no direct exchange between stations; all information exchange between stations must transit through the access point.

However this is a somewhat artificial concept as the AP is the only station which is connected to a distribution point (which often gives access to Internet) and if WPA or WPA2 authentication/encryption is used, each station can only communicate with the AP as only the AP knows how to decrypt the incoming frame.

In a managed wireless network the Access Point manages lists of associated STAs, STAs priority, power, as well as security policies.

To use AP mode in Linux you need to use "hostapd".

1.4.1.4 Service Set identifier (SSID)

An SSID (Service Set IDentifier) is a unique label that distinguishes a wireless local network from another. Therefore, all access points and stations trying to become part of a specific BSS must use the same SSID. The stations use the SSID for establishing and maintaining connectivity with access points.

The network is named after the MAC-Address (BSSID) of the AP. The human readable name for the network, the SSID, is also set by the AP.

1.4.1.5 Monitor (MON) mode

Monitor mode is often used in a passive mode, in this case no frames are transmitted. All incoming packets are handed over to the host computer completely unfiltered. This mode is useful to see what's going on the network.

With mac80211, it's also possible to transmit packets in monitor mode, which is known as packet injection or radiotap. This is useful for applications that wish to implement MLME work in userspace, for example to support nonstandard MAC extensions of IEEE 802.11.

Monitor mode interfaces always work on a "best effort" basis.

1.4.2 802.11 *INFRASTRUCTURES TYPES*

There are two classical modes of creating an 802.11 wireless LAN: "infrastructure mode and ad hoc mode." One corresponds to a mode where there is a point of administration of the LAN. The other mode is a basic point-to-point link between two terminals that are not specialized in some role: They are peers.

1.4.2.1 Mode infrastructure

This mode is called infrastructure mode because an 802.11 station will be able to use a set of services provided by the infrastructure of the network at an 802.11 access point. Communication between a station and another station is done via the access point (AP), this results in a division by two of the global throughput of the LAN. All stations belonging to the same BBS use the same radio channel. But communications are encrypted with a session key (WPA-2) that varies from one station to another and is valid for a single use. This allows a confidentiality of exchanges between station and point of access.

A coverage area is the geographic area in which an 802.11 station is able to interact with another 802.11 station. The basic coverage area is called a BSS (Basic Service Set)

Infrastructure mode supports communication in a one-to-many mode. There is only one single pair of nodes (an access point and a station) which can communicate simultaneously in an 802.11 BSS. However this situation changes with 802.11ac, because if the access point has multiple antennas, it may be able to fraction the BSS in several areas that are independent (MU-MIMO).

1.4.2.2 OBSS

This means a real life situation where multiple BSS are overlapping some geographic area (Overlapping Basic Service Sets). Nothing forbids several independent groups of stations to use the same media to exchange information. These groups overlap, but as they are not coordinated they interfere with each other. Indeed, the wireless spectrum being without license, a central coordinator which would allocate the access time to channel all of stations, would be very difficult to implement.

802.11aa was included in the 802.11 revision of the specification in mid-2013. 802.11aa brings a lot of features that would help stations to better cooperate in case of OBSS.

However, more and more a centralized controller is used to access to the radio media, especially in large business or entertainment areas.

1.4.2.3 PBBS, MU-MIMO, TLDS and ad hoc Mode:

802.11ad introduces a new mode that allows two stations to temporarily operate a private network to transfer information quickly from one station to another.

IEEE 802.11ac uses the technology of multi-user, multi-input multi-output (DL MU-MIMO) to transmit simultaneously from the access point to different stations through multiple spatial streams in contrast to the timesharing nature of 'Carrier Sense Multiple Access' (CSMA) currently used in 802.11 systems. This creates groups of stations inside the BSS.

A subset of the MAC functionalities is available for use between two 802.11ac stations that have a TDLS link. TDLS (Tunneled Direct Link Setup) is a technology that allows devices to automatically create a direct connection between them, avoiding the delays caused by congestion in the AP.

It is possible to use the ad-hoc mode to allow direct communication between stations. The coverage area of base in ad-hoc mode is then called IBSS (independent Basic Service Set).

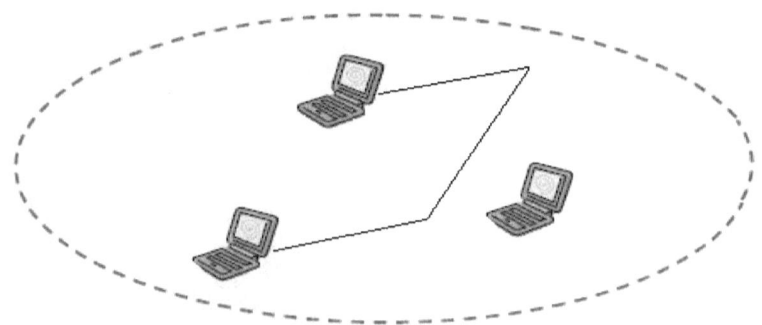

Figure 10 Ad hoc mode

This mode has recently been extended with the introduction of PBSS.

1.4.2.4 Basic Service Set Identification (BSSID)

The identification of basic services (BSSID) uniquely identifies each BSS. However, the SSID can be used in multiple BSS which can optionally overlap. In infrastructure BSS, the BSSID is the MAC address of the access point. In an IBSS (ad hoc mode), the BSSID is a locally administered MAC address that is generated from a 46 bit random number.

1.4.2.5 Extended Service Set

It is a set of one or more BSS interconnected and integrated that appear as a single BSS to the layer of logical link control (LLC) of any station associated with one of these BSS.

All of the interconnected BSS must be a common identifier (SSID) so stations can re-associate easily to another BSS. They can either work on the same channel, or work on different channels.

In the picture below two 802.11 LAN are connected by an Ethernet link between access points. This allows a station in the first BSS to access stations in the second BSS.

A station can migrate from BSS1 to the BBS2, because the means of authentication are shared.

The link between the two access points is called DS for 'distribution system '. It is important to make a distinction in one hand between two BSS connected by a DS that can communicate with each other, even though these two BSS have different SSID, on the other hand Extended Service Set that allows a station to migrate physically from one BSS to another BSS.

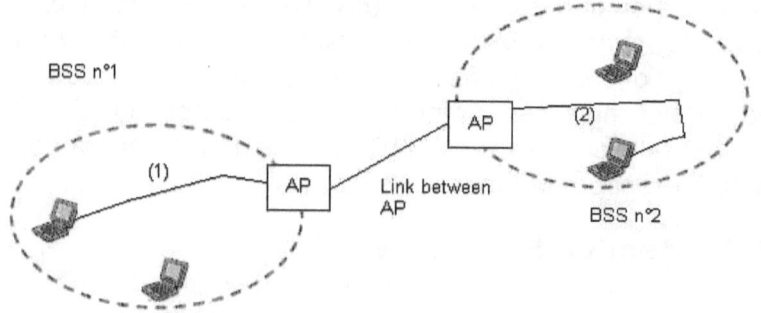

Figure 11 Extended Service Set

1.4.2.6 Mesh networks

It's a set of BSS where most of them do not have access to an access point and thus to the Internet. In this case the intermediate stations serve as relay station for those stations that wish to communicate with the access point.

2 LLC

IEEE 802.2 is the name given to the "Logical Link Control»
(LLC), which is the upper part of the OSI model data link layer.
The LLC sub layer presents a uniform interface to the software
using the data link service. The software using the data link
service is usually the IP layer and an application sits above the
IP layer. LLC allows different applications and different protocols
to simultaneously access a single network. LLC enables access
to a multiplicity of devices. Per se, LLC is not part of the 802.11
amendment but it helps to understand in which conceptual
framework, 802.11 is inserted. This will be very important in
future releases of the standard as interworking between 802.11
and other 802 technologies will occur more and more.

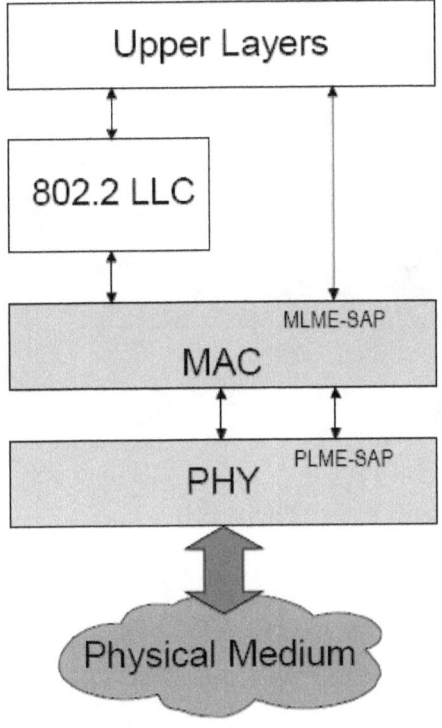

Figure 12 LLC in protocol stack

The description given here is only for the record. LLC manages the reception and transmission of data over a network on behalf of multiple client processes in the upper layer. For enabling a fair access time to each process the data are fractionated into packets to which LLC is adding information about the sending and the receiving process. LLC has to allow the exchange of data, between users of a LAN using the MAC 802 layer. The Media Access Control (MAC) sub layer, depends on the particular environment that is used (Ethernet, Token Ring, FDDI, 802.11, etc.), and in particular how it handles access contention. LLC can also make a link between two LANs. It is the layer that implements functions like "bridge", «hub", "router", "repeater" or "switch".

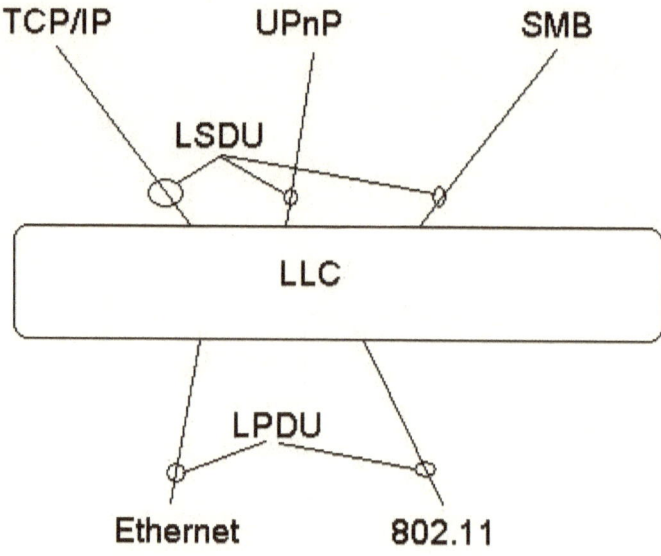

Figure 13 LLC as a hub between applications and devices

LLC operates as a multiple MAC Repeater that accepts MAC frames on input and route them on the appropriate output port.

There are also classes of services built from these types of services:

	Type I	Type II	Type III
Class 1	*		
Class 2	*	*	
Class 3	*		*
Class 4	*	*	*

These services apply to a communication between LLC peers using the LPDU format. "Logical Protocol Data Unit" is a message exchanged between peers LLC units.

"MAC Protocol Data Unit" which is shortened in "MPDU" is a message at the MAC-level between peer entities located on a LAN.

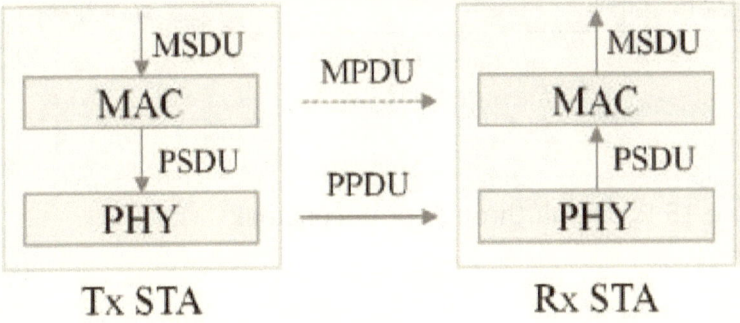

Figure 14 MSDU, MPDU and PPDU

It can be seen in the following figure how any given information, will be transferred by an application to the IP Protocol, which itself will use the services of the LLC layer. LLC then adds information about the sending process (the IP stack) as well as the information that identifies the recipient. Each protocol using LLC is identified by a "Service Access Point" (SAP). The LLC layer relies on the MAC layer for shared access to the media.

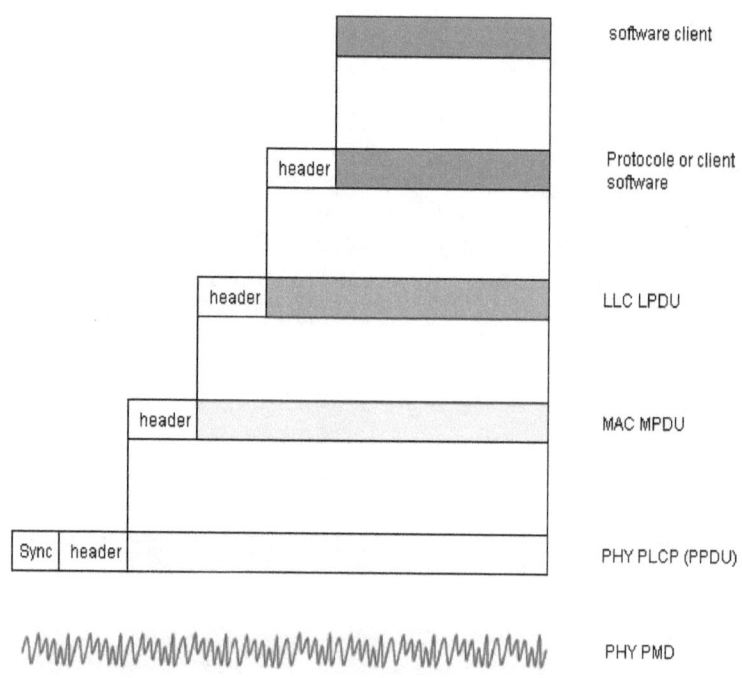

Figure 15 Data Unit through protocol stack

LLC provides three types of services to the next level (more on that later):

- Not connected, not acknowledged service (Type I)

- Not connected, acknowledged Service (Type II)

- Connected service (Type III)

The first type allows sending frames

- To a single destination (point to point or transfer unicast)

- To multiple destinations on the same network (multicast)

- Or all stations in the network (broadcast).

The use of multicasting and multicast enable reducing network traffic when the same information must be propagated to all stations of the network: It's only sent once to multiple recipients. However, the first type service gives no warranties concerning the order in which frames were received. The sender does not even get an acknowledgment that the frames have been received.

The second type is a connection-oriented mode. The numbering allows the frames received to be ordered as they were sent, even if they are received in another order, and acknowledgment makes it possible to check that all the frames have been correctly received. Errors in PDUs are also checked and reported. The error mechanism uses ARQ which has both positive and negative acknowledges. Negative acknowledges are sent only in case of error.

The third type is a service connection with acknowledge. There is no sequence control. It supports only peer-to-peer exchange. It uses positive acknowledge.

2.1 LLC HEADER

As seen above the 802.2 header includes two address fields of eight bits, called "Service Access Point" in OSI terminology, they include the destination (DSAP) and source (SSAP) fields. The low weight of the DSAP bit indicates if there is a single receiver or whether it is a group address.

If the low-order bit is 0, the remaining 7 bits of the address specify an individual address, which refers to a single point of local access service (LSAP) to which package is to be delivered.

If the low-order bit is equal to one, the remaining 7 bits of the address specify a group address, which refers to a LSAP group to which the package is to be delivered. The low weight of the SSAP bit indicates when it is equal to zero that the packet is a command packet, and if it is equal to one, that the package is a response packet. The remaining 7 bits of the SSAP specify the LSAP from which the packet was forwarded.

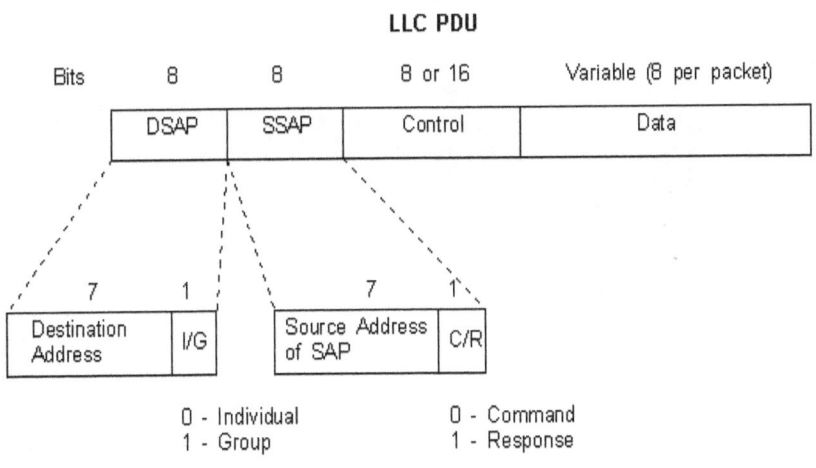

Figure 16 LLC PDU

Some values for the SSAP and DSAP are:

- 06: TCP/IP

- 7F: IEEE 802.2

- F0: Netbios...

LLC headers are important where the 802.11 MAC layer meets with other MAC layers such as 802.3 or even when there are inter-operating problems with the IP layer. In this later case it's worth investigating the Ethertype value but it's out of scope of this book.

2.2 LLC-MAC INTERFACE: MAC-SAP

2.2.1 INTRODUCTION

What is the difference between the LLC SAP and the origin and destination fields of an IP packet in one hand and the transmitter and recipient fields of a MAC frame in the other hand?

The origin and destination of an IP packet fields are used to identify the machine that uses the IP layer. IP is a protocol that establishes the conceptual equivalent of a straight cable between two machines (a socket) because equipment and software in charge of IP care about the routing between the two computers.

As IP must give the information sent by a higher layer to one or more routers, without knowing how it will be carried out or even if the routers have the ability to manage a transmission that may be quite long enough and may interfere with other transmissions by monopolizing a router, IP cuts the information in smaller packets and include in each packet, the identity of sender and recipient machines. IP is not concerned to know if the packets will arrive in the correct order, if one will be missing, or if one of them is corrupted. The reason for this great optimism by the IP layer is that it only deals with routing; its tries just to be independent of the notions of addresses on underlying different types of networks such as ATM/Ethernet, IPX, MPLS, X25.

Origin and destination of an LLC packet fields are used to identify the process that uses the LLC layer. As you remember, multiple processes can use the LLC layer at the same time. It is the role of LLC SAP fields to identify those processes.

The MAC layer must enable MAC entities to access a media that is shared between different machines within a local area network. Here in MAC the information is again divided in smaller packets. There again it raises the question of identifying the transmitter and the receiver of the package. There is no functional overlap with IP addresses as on the same machine there may exists several MAC interfaces, one for 802.11, the other for Ethernet, etc...

The 802.11 MAC model provides:
* The MAC-SAP interface to convey MSDUs from and to the LLC entity,
* The MLME-SAP and PLME-SAP interfaces use service primitives to receive and send indications, requests and confirmations from and to a station management entity,
* And the PHY-SAP and PMD-SAP interfaces use internal service primitives to interact with the PHY layer.

This book will study most of those interfaces.

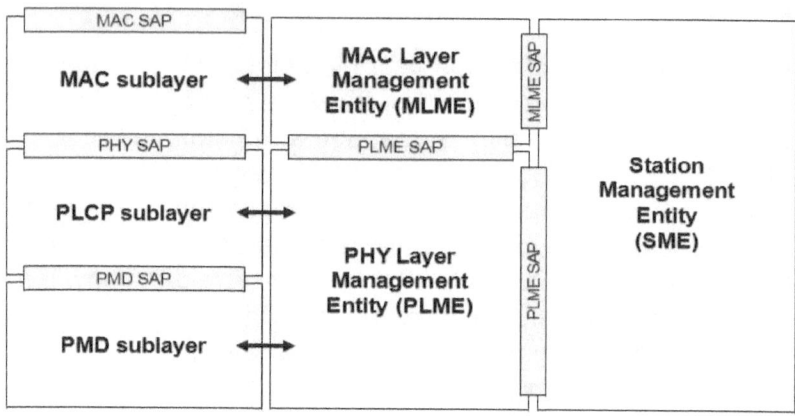

Figure 17 802.11 interfaces

2.2.2 LINUX LLC/MAC INTERFACES

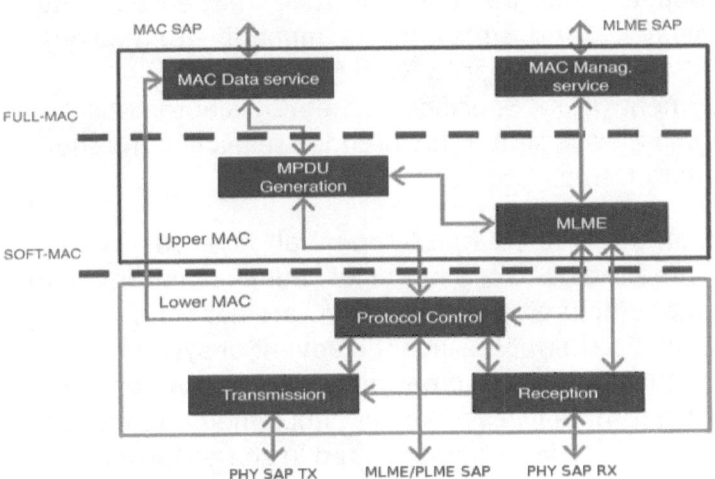

Figure 18: Linux SAP interfaces

Early 802.11 devices were designed according to a full-MAC approach. In those devices the MAC layer was almost entirely implemented in the card firmware, and only a small amount of code mainly implementing interfaces to the higher levels. The flexibility of commodity WLAN cards has significantly improved since a number of vendors (including Intel, Ralink. Realtek, Atheros, Broadcom), started to exploit an innovative soft-MAC design, by transferring to the host processor the non-time-critical MAC layer functions.

Still, even in the soft-MAC case, the "Lower MAC", which comprise crucial sub-systems such as transmission, reception and protocol control, remains hard-coded in the card. Although some chipsets (e.g. from Atheros and Broadcom) permit the tuning of selected MAC parameters (such as contention windows) via registers, more substantial MAC operation changes require access to the firmware code. some (time-critical) MAC functions cannot be changed, as they reside in proprietary firmware that is closely tied to the underlying Application-Specific Integrated Circuit (ASIC) that encapsulates the physical layer. For this reason, it is impossible to modify Request-to-Send (RTS), Clear-to-Send (CTS), and Acknowledgment (ACK). Specifically, the ACK is automatically generated and always sent if the intended receiver correctly receives a data frame.

There were attempts to create an "openHal" layer that would have eliminated the need of binary files. But even in the realm of a single manufacturer, the chipsets I/O are not compatible between them. Furthermore since the advent of system on chips "SoC", the layering becomes quite blurry. Another possibility to get more freedom in MAC implementation is to use the "monitor mode" that is implemented In some chipsets. This mode is often used to listen to frames as they are received by the PHY, sometimes it's even possible to send frames at the condition to re-implement the entire MAC layer.

Whenever the operating system has a packet to transmit, it will encapsulate it in a socket buffer and send it to the mac80211 driver. The socket buffer is an internal data structure of the network subsystem of the Linux kernel that represents a single packet. The mac80211 driver receives a link-layer packet with an 802.3 header that needs to be transformed into an 802.11 packet by the driver

The user space application makes use of the nl80211 calls to interact with cfg80211 which in turn communicates with mac80211. The controls are initiated from a user space application and are in turn transferred through system calls.

The wpa_supplicant uses nl80211 netlink functions

2.2.3 MAC-SAP PRIMITIVES
The 802.11 defines three primitives for the MAC-SAP: MA-UNITDATA.request, MA-UNITDATA.indication and MA-

UNITDATA.confirm. These primitives transport data. The transport of these data units is performed in an asynchronous manner.

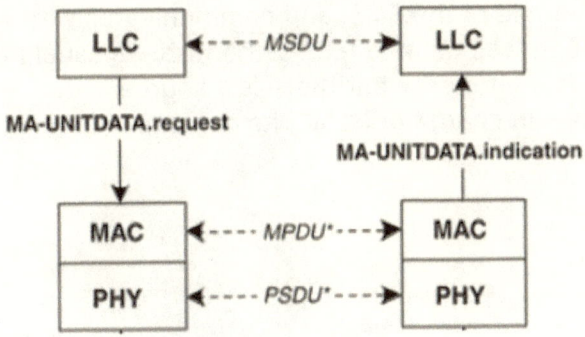

Figure 19 MA-UNITDATA

- The LLC layer requests a transmission of an MSDU by the primitive "MA-UNITDATA.request".
- The MAC layer generates a primitive "MA-UNITDATA.confirm" to the LLC after transmission (or failure of transmission) of an MSDU.

- When the station receives an MSDU via the radio media, the MAC layer generates a primitive "MA-UNITDATA.indication" to the LLC.
- The MAC layer also generates a primitive "MA-UNITDATA.indication" to the LLC about some status information of local significance.

This primitive belongs in fact to LLC.

2.3 Management layer

The user must be able to choose when an to which AP to connect. Usually he uses an application to do so. This application plays the role of the SME, and communicates with the MAC layer through well-defined IEEE semantics. As usual in Linux the situation is a bit messy but there is a source code file in mac80211 which is in charge of MLME, its name is indeed "mlme.c".

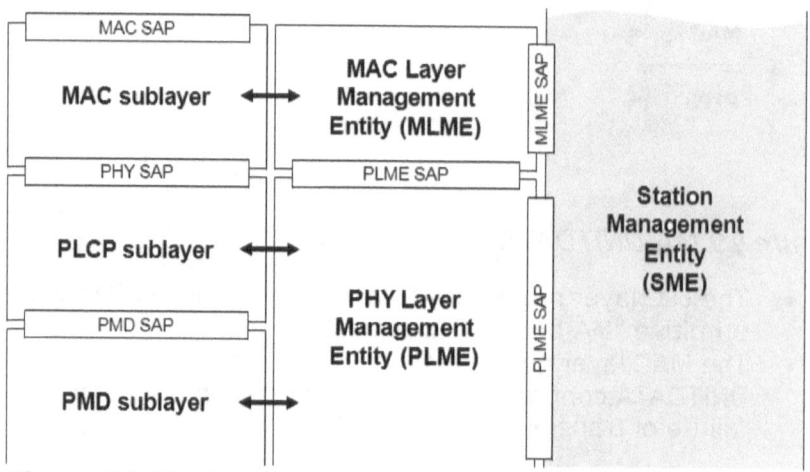

Figure 20 Station Management Entity (SME)

2.2.4 SME

In order to ensure the proper functioning of the MAC layer, a station management entity (SME) is present in each station. The SME is an independent entity of the layer that sits in a separate management plan. In a PC or mobile phone it's the software that interacts with the user to present him a list of networks, receive the passphrase and manage disconnection and reconnection. It's neither an application nor a feature of an 802.11 station, it's a feature of the operating system that manages the application and it's relation to the 802.11 station. Some of these interactions are defined explicitly in the 802.11ac, via a standard SAP. This definition includes GET and SET between MLME, PLME and SME as well as other service primitives. The GET and SET act on information that is called "objects" in 802.11 literature. Those objects are often states of

the 802.11 seen as a state machine. It's modeled after SNMP objects.
Other interactions are not defined explicitly in the 802.11ac standard, for example the interfaces between the MAC and the MLME layer and between the PLME / PLCP and PMD, or the precise manner in which these SME, MAC and PHY are integrated into the MAC and PHY layers is not specified in the standard 802.11ac.

The Service Access Point in this model is the following:
- SME- MLME SAP
- SME-PLME SAP
- MLME-PLME SAP

The last two SAP support identical primitives and actually could be considered as a single set that is used either directly by the MLME SMEs.

2.2.5 MLME SAP, GETTING AND SETTING MIB VALUES

GET and SET primitives are represented as queries. These primitives live in MLME or PLME interfaces so their prefix depends on whether the MAC or PHY layer is involved. For example a GET request at MAC level is "MLME-GET.request (MIBattribute)" and a GET request at PHY level is "PLME-GET.request (MIBattribute)".

They are used to read or write the MIB values. MIB values represent states or status information of the MAC or PHY layer. Sometimes they are called the "native 802.11 data types". A management information base (MIB) is a virtual database used for managing the entities in a communications network. Those entities names begin with dot11...

However there is no notion of 802.11 MIB in Linux. A utility makes it possible to map some 802.11 information on SNMP protocol to make it possible to manage 802.11 hardware with SNMP software.

It's the only formal thing in the 802.11 specification. Many actions or parameters are described in relation to dot11 elements. Objects in the MIB are defined using a subset of Abstract Syntax Notation One.

MLME-GET.request (MIBattribute)
Request the value of the MIBattribute data.

MLME-GET.confirm (status, MIBattribute, MIBattributevalue)
This primitive returns the appropriate attribute MIB if successful; otherwise it returns an error indicator in the status field. Possible error status values include "invalid attribute MIB" and "trying to read, a write-only attribute MIB."

MLME-SET.request (MIBattribute, MIBattributevalue)
These primitive requests the specified MIB attribute to be set to the given value. If this MIBattribute parameter involves a specific action, it demands that the action is performed.

MLME-SET.confirm (status MIBattribute)
If the Status field is set to "success", this confirms that the MIB attribute has been set to the requested value. Otherwise, this primitive returns an error indicator in the status field. If the

MIBattribute parameter involves a specific action, this confirms that the action has been performed. Possible error status values include "invalid attribute MIB" and "attempting to write, the read only MIB attribute."

In addition, there are some applications that can be invoked in a SAP data, which do not uses any parameter or to obtain a specific attribute of the MIB.

MLME-RESET.request:

This service is used to initialize the management entities, and the MIB. It could include a list of attributes for the elements to be initialized to the default values.

2.2.6 OTHER MLME SAP INTERFACE

MLME SAP interface makes it possible for an external application (the SME) to make specific request to the MAC, for example, searching BSS in the vicinity, associate with one of them, etc. Those messages give only the semantic, in a given implementation the real API might use a very different syntax.

The following list is only partial and is intended to give an idea of what is proposed.

The MLME SAP primitives follow a general pattern such as ACTION.request is followed by ACTION.confirm (for an Exchange initiated by the SAP client, i.e. the SME) and ACTION.indication is followed by ACTION.response (for an Exchange initiated by the MLME).

For example for making an 802.11 station to scan the radio channels to search for an access point beacon, here is how it goes:

1. The SME sends a MLME-SCAN.request to the MLME SAP interface; the parameters indicate either a SSID to search for, or a broadcast SSID, which means that every beacon frame will be OK.

2. The MLME-SCAN.confirm returns information about the beacon frames it found, in the BSSDescriptionSet parameter.

Here is an example with parameters. This primitive request association with a MAC-level peer located to one access point:

```
MLME-ASSOCIATE.request (
PeerSTAAddress,
AssociateFailureTimeout,
CapabilityInformation,
ListenInterval,
Supported Channels,
RSN,
```

```
QoSCapability,

Content of FT Authentication elements,

SupportedOperatingClasses,

HT Capabilities,

Extended Capabilities,

20/40 BSS Coexistence,

QoSTrafficCapability,

TIMBroadcastRequest,

EmergencyServices,

DMG Capabilities (11ad)

Multi-band local (11ad)

Multi-band peer (11ad)

Multiple MAC Addresses (11ad)

VHT Capabilities,

VendorSpecificInfo

MLME-POWERMGT.request
)
```

2.2.6.1 About starting/stopping a new BSS:

MLME-START.request asks the MAC layer to start a new BSS or join a SBMS.

MLME-START.confirm reports the results of a BSS creation or becoming a member of a SBMS.

MLME-STOP.request requests that the MAC layer stops a BSS previously started using a MLME-START.request primitive.

MLME-POWERMGT.confirm requires a change in power management mode.

MLME-SCAN.request request a scanning of radio channels in search of a beacon (passive scanning).

2.2.6.2 Search an access point and associate:

MLME-SCAN.confirm initializes a search for BSSs nearby.

MLME-JOIN.request returns the descriptions of all of the BSSs detected by the research process.

MLME-JOIN.confirm asked the association of this station to a BSS.

MLME-AUTHENTICATE.request confirms the integration in a BSS.

MLME-AUTHENTICATE.confirm requests authentication with a peer at the MAC level.

MLME-AUTHENTICATE.indication reports results from an attempt to authenticate with a peer at the MAC level.

MLME-AUTHENTICATE.response indicates receipt of an authentication request by a peer at the MAC level by the station dealing with this primitive.

MLME-ASSOCIATE.indication reports result from an attempt to associate with an access point.

MLME-ASSOCIATE.response reports the answer to a request to associate with the local MAC entity of an access point.

MLME-REASSOCIATE.request is used for an access point, to send a response to a peer at MAC-level who asked an association.

MLME-REASSOCIATE.confirm requests a change to the association with an access point.

MLME-REASSOCIATE.indication reports result of a change to the association with an access point.

MLME-REASSOCIATE.response indicates a response in a change to the association with an access point.

2.2.6.3 Leave an access point BSS:

MLME-DEAUTHENTICATE.request is used to send a response to a MAC-level peer which had asked to be authenticated by the station that received this primitive.

MLME-DEAUTHENTICATE.confirm asks for a relationship with a MAC-level peer to be invalidated:

MLME-DEAUTHENTICATE.indication reports results from an attempt to de-authenticate with a peer at MAC-level:

MLME-DISASSOCIATE.request requires a dis-association with a pair at MAC-level.

MLME-DISASSOCIATE.confirm reports the results of a dis-association with another MAC entity.

MLME-DISASSOCIATE.indication reports about the dis-association with a peer entity at the MAC-level.

MLME-RESET.request requests that the MAC layer is reset.

2.2.7 PLME SAP INTERFACE

The management of the PHY service interface consists of the primitives PLME-GET and PLME-SET, those are similar to their MAC SAP counterpart as described previously.

They consist of *PLME-GET*.request (MIBattribute), *PLME-GET*.confirm (status, MIBattribute, MIBattributevalue), *PLME-SET*.request (MIBattribute, MIBattributevalue), *PLME-SET*.confirm (status MIBattribute), *and PLME-RESET*.request.

There are also the following primitives PLME-RESET and PLME-CHARACTERISTICS.

PLME-RESET.request
This primitive is a request by the SME to reset the PHY. The PHY layer always comes back to the "receiving" State in order to avoid the inadvertent transmission of data.

PLME-CHARACTERISTICS.request
This primitive is a request made to the SME to provide the operational characteristics of PHY.

PLME-CHARACTERISTICS.confirm
This primitive provides PHY operating parameters.

2.2.8 LINUX MLME

A standard wireless driver with Linux wireless capabilities includes some kernel modules and provides interfaces used by user level tools to configure the device behavior. The main modules defined in the framework are the mac80211 and the cfg80211; these modules are loaded and used by the drivers (e.g., ath5k, ath9k, b43) that are implemented in separate Linux kernel modules.

The mac80211 is nevertheless a difficult to understand set of sub-modules highly interconnected, e.g. : the MAC layer management entity *(mlme)* the high throughput or the MPDU aggregation are not implemented as separated modules, thus preventing any kind of modularization.

3 MAC LAYER

3.1 INTRODUCTION
The 802 data link layer is divided into two sub layers:

- The logical link control layer (Logical Link Control, noted LLC) and its interface to the MAC layer, was seen in the previous chapter

- The access control layer support (Media Access Control or MAC), which is the subject of this chapter

The role of the MAC layer is to implement communication means between peer LLC entities. In the normal operating mode, the transport of an MSDU is done at "best effort"; it means that there is no guarantee of success. However in mode "Quality of Service", a traffic identifier (TID) in the MSDU is used to specify different service levels. This mode also allows a more synchronous operation similar to a BSD "pipe" or a TCP socket.

One difficulty in 802.11 is that the radio media is shared between different devices which makes it particularly difficult to manage access contention (access competition), and this has influenced the design of the 802.11 MAC layer.

3.1.1 INTERFERENCE, HIDDEN STATION.
3.1.1.1 adjacent channel interference
Cells can overlap, which can potentially create interference problems.

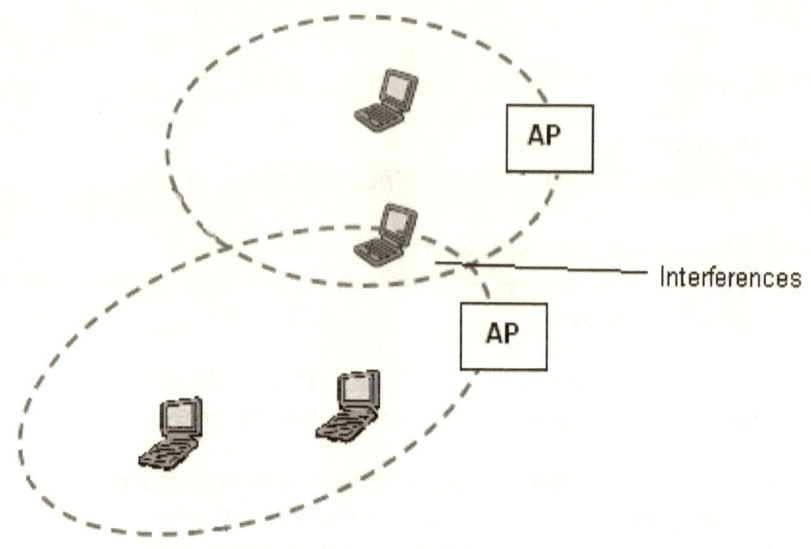

Interferences

Figure 21 Adjacent channel interference

To remedy to this problem we can use different channels for the two BSS. However there is a significant interference between adjacent channels and the number of available channels may be limited by regulation and by the type of 802.11 network. Mitigating interference is discussed in several paragraphs in this book.

3.1.1.2 Hidden station

Figure 22 Hidden station

A problem inherent to sharing a media, is that of deafness to other stations. The simplest case is that of the station which is out of range of another, with both in the access point reach. The two stations may emit at the same time because each are unaware that the other emits. Fortunately this issue can be settled with a media access protocol which is managed by the access point.

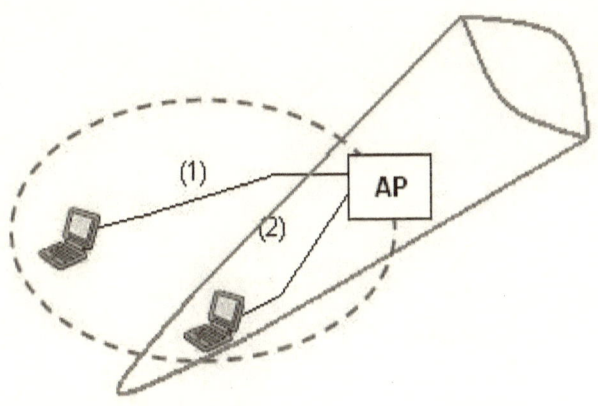

Figure 23 Directional antennas

Another problem is the use of directional antennas, it can make, not only the station (2) to the station (1) but also make the point of access deaf to station (1).

The MAC layer makes it possible to provide access to the physical media, to join a network and to provide authentication and privacy services. It transforms the activity on the radio channel into digital frames that could be processed by computers with such mathematical techniques such as FFT, LDPC coding or other matrix transformations.

The 802.11 MAC layer is comparable to the 802.3 MAC layer: it implements the policy for accessing to the radio media.

However, this MAC layer is IEEE 802.11-specific because it offers more functions than a classic MAC layer (allocation of media, addressing, frames formatting). For example there are specific functions for association with a network, authentication and confidentiality through encryption.

3.1.2 SESSION LIFE CYCLE

This is about a full life cycle of a station 802.11, from power up, its integration into a Wi-Fi cell and its subsequent withdrawal. In a system of shared access to the radio media, it is important that a message can block others, also exchanges take place by packets of information to allow everyone access to the media when it is clearly free.

3.1.2.1 Making contact

When a station is switched on, its MAC layer is made running; the software implementing the LLC layer above the MAC layer must establish an association with an access point by scanning the different radio channels to detect networks that emit "beacon" frames. A beacon frame is emitted at regular intervals by an access point and it provides different information to the station enabling it eventually to attempt an association with the access point. The scanning can be active or passive. The IEEE specification allows different implementations. The MAC layer relies on the PHY layer to translate digital information into radio signals and vice versa.

The access point periodically sends a beacon frame to announce its presence and relay information such as the time stamp, SSID, and other parameters.

In Linux beacon management at the AP, is achieved with the following calls to functions:

```
static int ieee80211_assign_beacon(struct
ieee80211_sub_if_data *sdata, struct
cfg80211_beacon_data *params)

static int ieee80211_change_beacon(struct wiphy
*wiphy, struct net_device *dev, struct
cfg80211_beacon_data *params)
```

In IEEE 802.11 specification this is done with:

MLME-START.request asks the MAC layer to start a new BSS or join a SBMS.

MLME-START.confirm reports the results of a BSS creation or becoming a member of a SBMS.

MLME-STOP.request requests that the MAC layer stops a BSS previously started using a MLME-START.request primitive.

3.1.2.1.1 Active scanning

The station seeking to establish contact with an access point, sends a "Probe Request' on a free channel without waiting to know for the identity of the access point. Active Scan is in effect, the fastest way to search for an access point, but it consumes more battery power. Indeed emitting radio waves is costly in power.

Active Scan consists to emit a frame on the access point radio channel. This is to prevent having to listen to each channel of the access point to discover 'beacon' frames, which are sent with a long periodicity (100ms) by the access point. If Service Set identity matches that of the recipient field, the queried access point sends a response to the station that sent a 'Probe Request. The station uses this information to decide whether or not it will join in the BSS.

3.1.2.1.2 Information request frame (probe request)

A station sends a request frame for information when necessary to another station. For example, a station can send a request for information to determine what access points are within range.

```
void ieee80211_send_probe_req(struct
ieee80211_sub_if_data *sdata, u8 *dst,
        const u8 *ssid, size_t ssid_len,
        const u8 *ie, size_t ie_len,
```

```
        u32 ratemask, bool directed, u32 tx_flags,
        struct ieee80211_channel *channel, bool
scan)
```

3.1.2.1.3 Probe response frame

A station will respond to a request frame for information by a response frame containing the requested information.

static int **ieee80211_set_probe_resp**(struct ieee80211_sub_if_data *sdata, const u8 *resp, size_t resp_len)

3.1.2.1.4 Passive scanning

In the passive scan mechanism a station searches beacon frames on each channel of a cell to find the identity of the BSS: Its SSID. This mechanism can be slow, especially as it has to start over for each cell that would be accessible from the station. After selecting a channel, the station begins to listen for beacon frames which are circulated periodically. A beacon frame sent from an access point in a BSS or ESS contains information about the BSS and a timing reference. The transmission of beacon frames occurs every 100 ms.

As an example the two next MLME message semantics are used for querying a channel scan. In Linux those semantics are conveyed by cfg80211 functions.

MLME-SCAN.request request a scanning of radio channels in search of a beacon (passive scanning).

MLME-SCAN.confirm initializes a search for BSSs nearby.

The scanning process itself is fairly simple, but cfg80211 offers quite a bit of helper functionality. To start a scan, the scan operation will be invoked with a scan definition. This scan definition contains the channels to scan, and the SSIDs to send probe requests for (including the wildcard, if desired). A passive scan is indicated by having no SSIDs to probe. Additionally, a scan request may contain extra information elements that should be added to the probe request. The IEs are guaranteed to be well-formed, and will not exceed the maximum length the driver advertised in the wiphy structure.

When scanning finds a BSS, cfg80211 needs to be notified of that, because it is responsible for maintaining the BSS list; the driver should not maintain a list itself. For this notification, various functions exist.

Since drivers do not maintain a BSS list, there are also a number of functions to search for a BSS and obtain information about it from the BSS structure cfg80211 maintains. The BSS list is also made available to userspace.

Some device manufacturers that provide a MAC layer design their functions names with a name similar to the 802.11 name. Alas this is not the case in Linux, making the Linux 802.11 stack learning curve quite steep.

Here are the Linux functions for scanning channels in search of a BSS.

```
/**
 * notify that scan finished
 *
 * @request: the corresponding scan request
 * @aborted: set to true if the scan was aborted for
any reason, userspace will be notified of that.
 */

void cfg80211_scan_done(struct cfg80211_scan_request
*request, bool aborted);

/**
 * notify that new scan results are available
 *
 * @wiphy: the wiphy which got scheduled scan
results
 */
```

```c
void cfg80211_sched_scan_results(struct wiphy
*wiphy);


/**
 * notify that the scheduled scan has stopped
 *
 * @wiphy: the wiphy on which the scheduled scan
stopped
 *

 * The driver can call this function to inform
cfg80211 that the scheduled scan had to be stopped,
for whatever reason. The driver is then called back
via the sched_scan_stop operation when done.
 */

void cfg80211_sched_scan_stopped(struct wiphy
*wiphy);


/**
 * inform cfg80211 of a received BSS frame
 *
 * @wiphy: the wiphy reporting the BSS
 * @channel: The channel the frame was received on
 * @mgmt: the management frame (probe response or
beacon)
 * @len: length of the management frame
 * @signal: the signal strength, type depends on the
wiphy's signal_type
 * @gfp: context flags
 *
 */

struct cfg80211_bss * __must_check

cfg80211_inform_bss_frame(struct wiphy *wiphy,
            struct ieee80211_channel *channel,
```

```
            struct ieee80211_mgmt *mgmt, size_t len,
            s32 signal, gfp_t gfp);

/**
 * This informs cfg80211 that BSS information was
found
 * @wiphy: the wiphy reporting the BSS

 * @channel: The channel the frame was received on
 * @bssid: the BSSID of the BSS
 * @tsf: the TSF sent by the peer in the
beacon/probe response (or 0)
 * @capability: the capability field sent by the
peer
 * @beacon_interval: the beacon interval announced
by the peer
 * @ie: additional IEs sent by the peer
 * @ielen: length of the additional IEs
 * @signal: the signal strength, type depends on the
wiphy's signal_type

 * @gfp: context flags
 *
 */

struct cfg80211_bss * __must_check

cfg80211_inform_bss(struct wiphy *wiphy,
        struct ieee80211_channel *channel,
        const u8 *bssid, u64 tsf, u16 capability,
        u16 beacon_interval, const u8 *ie, size_t
ielen,
        s32 signal, gfp_t gfp);
```

cfg80211_inform_bss() is the function bearing the **MLME-SCAN**.indication semantics.

3.1.2.2 Authentication

Before any user data can be transferred, it requires that the sender station and the access point agree to exchange information. The two parties will be ready to exchange information when they will be sure of the identity of the other party. It is a prerequisite for an association. Over the amendments there were several versions of authentication protocols. Generally, this implies that the station and access point, either have a shared secret or they have inalienable electronic certificates that have been provided by third parties (Certificate Authorities). In both cases the part that wants to ensure the identity of the other Party asked it to encrypt a text with the secret, and checks the capability of the other party to provide a correct answer.

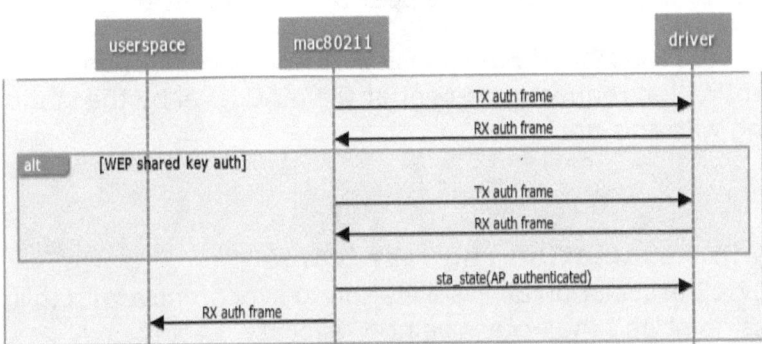

Figure 24: WEP authentication in Linux

In Linux this is initiated by a call to:

```
int cfg80211_mlme_auth(struct
cfg80211_registered_device *rdev struct net_device
*dev, struct ieee80211_channel *chan,   enum
nl80211_auth_type auth_type, const u8 *bssid,
const u8 *ssid, int ssid_len, const u8 *ie, int
ie_len, const u8 *key, int key_len, int key_idx,
const u8 *sae_data, int sae_data_len)
```

In IEEE 802.11 this is the semantics used:

MLME-JOIN.request returns the descriptions of all of the BSSs detected by the research process.

MLME-JOIN.confirm asked the association of this station to a BSS.

MLME-AUTHENTICATE.request confirms the integration in a BSS.

MLME-AUTHENTICATE.confirm requests authentication with a peer at the MAC level.

MLME-AUTHENTICATE.indication reports results from an attempt to authenticate with a peer at the MAC level.

MLME-AUTHENTICATE.response indicates receipt of an authentication request by a peer at the MAC level by the station dealing with this primitive.

3.1.2.3 Association request frame
An 802.11 association allows a station to synchronize with this access point and the access point to allocate resources. A station begins the association process by sending an application for association to an access point. This frame includes information about the station (for example, the throughput it can support) and the SSID of the network to which it wishes to associate itself. After having received the request for association, the access point studied it and if accepted, it reserves memory space and establishes an Association IDentifier for the station.

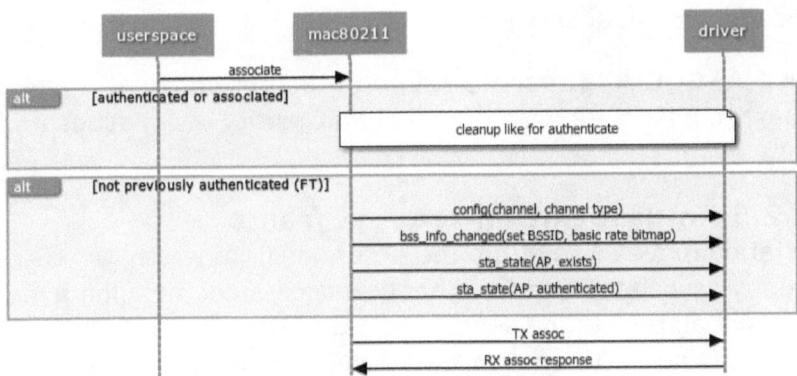

Figure 26: Authentication in Linux

In Linux this is initiated by a call to:

int **cfg80211_mlme_assoc**(struct
cfg80211_registered_device *rdev,
        struct net_device *dev,
        struct ieee80211_channel *chan,
        const u8 *bssid,
        const u8 *ssid, int ssid_len,
        struct cfg80211_assoc_request *req)

MLME-ASSOCIATE.request is the query for associating by a station to an AP.

3.1.2.4 Frame response to an association request

The access point sends a response frame with the association acceptance or a notice of rejection to the station that made the request. If the access point supports the association with the station, the frame includes information about the association, such as the data rates supported. If the association is positive, the station can use the access point to communicate with other stations on the network.

101

MLME-ASSOCIATE.indication reports result from an attempt to associate with an access point.

MLME-ASSOCIATE.response reports the answer to a request to associate with the local MAC entity of an access point.

3.1.2.5 Re-association request frame

If a station moves away from the access point to which it is currently associated and finds another access point having a better signal, the station sends a re-association frame to the new access point. The new access point will then coordinate the transmission of data frames which may still be waiting in the buffer of the previous access point.

3.1.2.6 Re-association response frame

The access point sends a response frame to a station applying for re-association. The response frame contains a notice of acceptance or rejection of the re-association request. Similarly to the association process, the frame response includes information about the association as association ID and supported throughput.

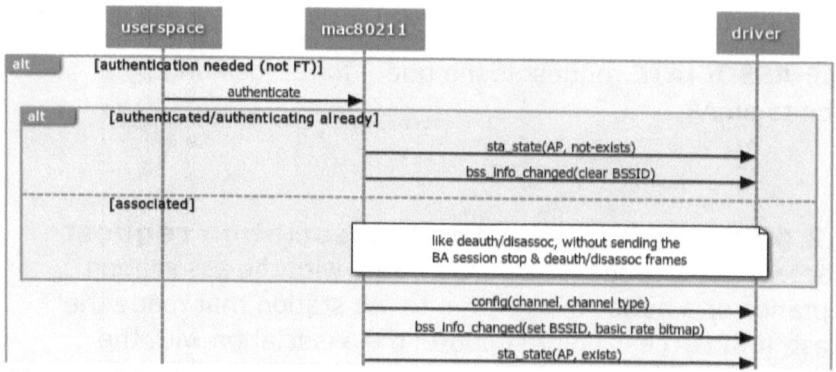

Figure 28: Association in Linux

3.1.2.7 Association end

There are also two possible methods for a station to end an association with a BSS.

- A rather elegant one consists for a station to explicitly request the end of the association. For example, a station that is shutting down gracefully can send a dissociation frame to alert the access point that the station is turned off. The access point can then remove memory allocations and the station from the association table.

- The other is to no longer interact with the AP for example because that station has been turned off. In this latter case the association is broken by the AP after a while when it sees no activity since some time.

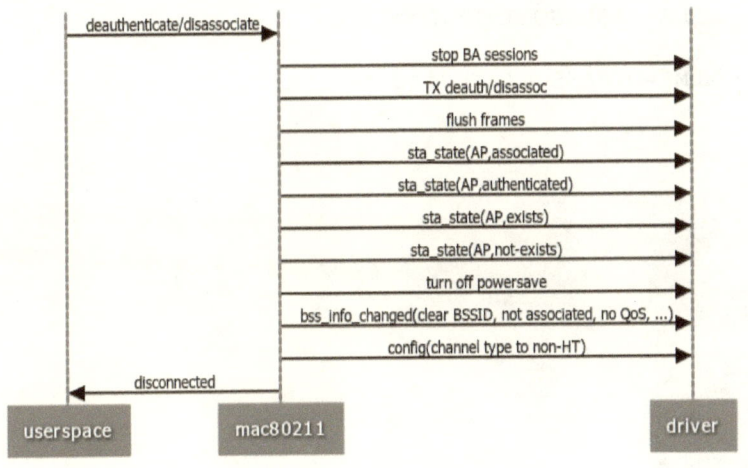

Figure 30 End of association in Linux

In Linux this is done with:

```
void cfg80211_send_disassoc(struct net_device
*dev, const u8 *buf, size_t len)

int

cfg80211_mlme_disassoc(struct
cfg80211_registered_device *rdev,

  struct net_device *dev, const u8 *bssid,

  const u8 *ie, int ie_len, u16 reason,

  bool local_state_change)
```

In IEEE 802.11 the semantics are specified with:

```
MLME-DISASSOCIATE.request

MLME-DISASSOCIATE.confirm

MLME-DISASSOCIATE.indication
```

3.1.3 RECENT EVOLUTIONS OF THE MAC LAYER

The MAC layer has been further developed since 802.11n to increase efficiency through frames aggregation and features enrichment.

In addition to the standard BSS mode where an AP coordinates the traffic of a group of stations, two 802.11ac stations may be members of an IBSS (ad-hoc connection mode between two stations, that is better known under the name of Wi-Fi direct). Some other features of the MAC layer are variations on the IBSS, like PBSS and so on.

A subset of the MAC functionalities is available for use between two 802.11ac stations that have a TDLS link. TDLS (Tunneled Direct Link Setup) is a technology that allows devices to automatically create a direct connection between them, avoiding the delays caused by congestion in the AP.

Similarly, a subset of the MAC functionality is available for use between two 802.11ac stations which have established links of "Mesh" type (connection with a network of stations without central coordination).

MIMO (a PHY feature) is done by exploiting the radio paths thanks to sounding and reporting frames which are MAC features.

Beam forming has impact on frame acknowledgment.

Multi-user operation, between an 802.11ac access point and several 802.11ac stations, is possible in a BSS and needs special MAC adaptation. 802.11ac allows the optional use of DL-MU-MIMO. DL-MU-MIMO makes it possible to create up to four A-MPDUs each supporting MPDU for as many associated multi-user stations. MU-MIMO has also impact on frame acknowledgment.

The access point uses the IDs Group (GID) to communicate with the stations involved in multi-user communication. Transmission of a single, multi-user PPDU provides a way to increase the overall throughput if compared to what would be obtained by sending an A-MPDU in several single user PPDU.

3.1.4 MAC LAYER ROLE

In a classical Ethernet local network where multiple hosts are communicating on a single media, the access method used by the machines is CSMA/CD (Carrier Sense Multiple Access with Collision Detection). Each machine that wants to send a message, checks that no other message is sent at the same time by another machine. If this is the case, the two machines wait for a random time before going back to the sending process.

In a radio environment this process is not possible because two stations transmitting simultaneously are not able to detect that another transmission occurs on the same channel. Thus the 802.11 standard offers a protocol quite similar to CSMA/CD but with some modifications, called CSMA/CA (Carrier Sense Multiple Access with Collision Avoidance).

The role of the MAC sub layer is primarily to:

- Define the frames headers by inserting information appropriately, so that the receiver can determine the purpose of the frame.

- Detect transmission errors, for example by using a checksum inserted by sender and checked by the receiver.

- Insert the source and destination MAC addresses in each transmitted frame.

- Filter frames received, looking only for those that sent to itself, by checking the destination MAC address.

- Control access to a shared physical media and keeping information about the free/busy state.

- Manage the relationship with the access point (synchronization, authentication, and encryption).

- Manage flow control, including aggregated packets (A-MSDU)

- Manage radio channel characteristics discovery through sounding and reporting frames

The downside of this method of access is that it is probabilistic: it is not possible to guarantee a minimum time before the media access becomes free, which is problematic for some applications (voice, video...). See 802.11e, which was incorporated in 2012 802.11 standard, for more sophisticated mechanisms.

3.1.6 MAC COMPONENTS

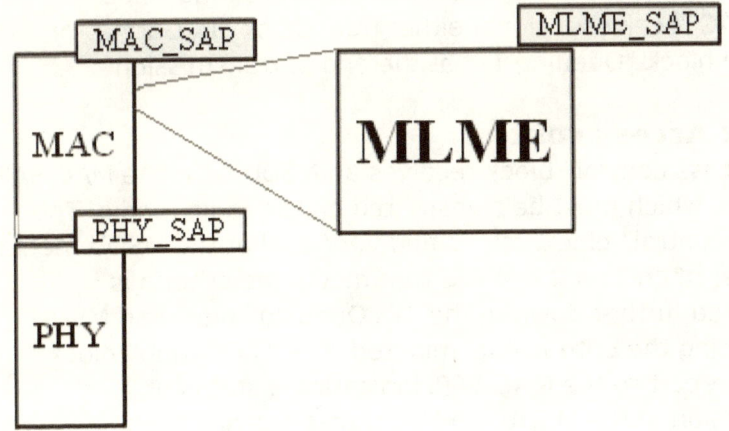

Figure 31 Mac interfaces

The following table shows the main blocks in the MAC layer.

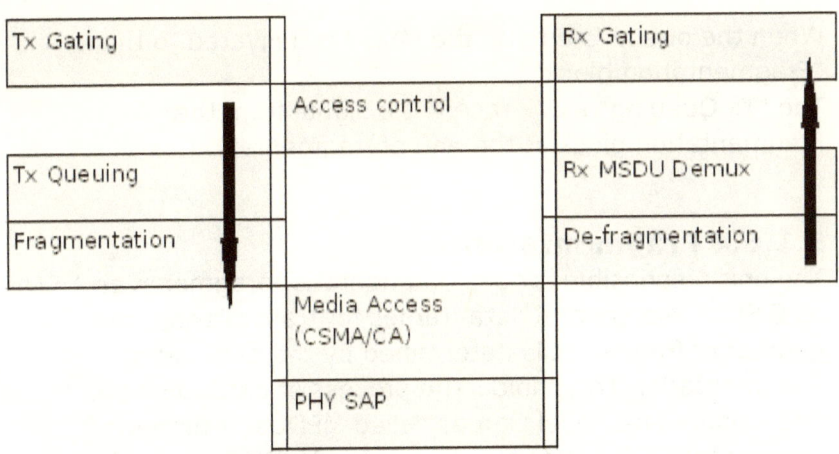

Figure 32 MAC components

3.1.6.1 MAC Tx Gating blocks

The MAC-SAP transfers an MSDU to the "Tx Gating" block.
SAP - MAC receives a signal either from block "Tx Gating" or
from the block "Queuing Tx" at the end of transmission

3.1.6.2 Access control

The "Access control" block receives an MSDU from the MAC-SAP
interface, which must be transmitted on the radio media. The
"Access control" block, stores and verifies that the MSDU meets
a number of criteria; an MSDU that meets the criteria is
transferred further down to the "Tx Queuing" block; an MSDU
not meeting the criteria is eliminated. The "Tx Gating" block
sends a report to the MAC-SAP, indicating a status of
transmission of the MSDU. an MSDU that has not been
transmitted is deleted internally (there is no retransmission,
and it doesn't bloc another MSDU.

3.1.6.3 Tx Queuing

The "Tx Queuing" block accepts an MSDU from the Tx Gating
block, it will queue it prior to the actual transmission. The "Tx
Queuing" Block reports a status of the transmission of the
MSDU to the MAC-SAP.

When the queue is empty, the MSDU is delivered to the
"Fragmentation block".
The "Tx Queuing" block receives a signal from the
Fragmentation block at the end of fragmentation.

3.1.6.4 Fragmentation

The unit responsible for the fragmentation fragments an MSDU
(LLC SDU) into several data frames for transmission, the
number of fragments is determined by the parameter
'Fragmentation Threshold'. The process of partitioning an MSDU
into smaller MAC level frames called MPDUs, is named
fragmentation. Fragmentation creates MPDUs smaller than the
original MSDU length to increase the probability of successful
transmission of the MSDU. MAC may use fragmentation to use

the medium efficiently (for example if a contention free time is granted through a TXOP). The inverse process of recombining MPDUs into a single MSDU is called "defragmentation".

The fragmentation block, forwards data frames to the "Media access" block.
It receives a signal from the «Media access" block at the end of the request processing about the status of transmission of the data frame.
The fragmentation block, signals the transmission to the MSDU to the "Tx Queuing" block.

In modern high throughput, the relative time devoted to transmit a data of constant size in the PHY frame is diminishing as the preamble is still mostly sent at a low throughput. Now the preamble may use most of the PHY frame duration if nothing is done to maximize the amount in data in an MPDU. So in a counter-intuitive movement in 802.11ac, MPDUs that are sent to the same station are transported in a single huge MPDU named A-MPDU, which itself uses a single MU VHT PPDU. Even an aggregated MPDU, hasn't a longer duration than the original MPDU, so the original goal is still achieved.

An MSDU transmitted within an A-MPDU that does not contain a VHT single MPDU shall not be fragmented even if its length exceeds dot11FragmentationThreshold. Group addressed MSDUs shall not be fragmented even if their length exceeds dot11FragmentationThreshold.

When an individually addressed MSDU is received from the LLC that would result in an MPDU of length greater than dot11FragmentationThreshold, the MSDU shall be fragmented. Each fragment is a frame no longer than dot11FragmentationThreshold.

The MPDUs resulting from the fragmentation of an MSDU are sent as independent transmissions, each of which is separately acknowledged. This permits transmission retries to occur per fragment, rather than per MSDU. The fragments of a single MSDU are either

- Sent during a Contention Period as individual frames using the DCF, or

- Sent during a Contention Free Period as individual frames obeying the rules of the PC medium access procedure, or

- Sent as a burst in an EDCA or HCCA TXOP, subject to TXOP limits.

3.1.6.5 Media Access

It accepts MPDUs (fragmented LLC MSDUs) for transmission, from the Fragmentation bloc. It is seeking access to the radio media and (if possible) passes the MPDUs as a sequence of PSDUs. When this task is completed it passes back status information to the Fragmentation block. It accepts reports from PHY-SAP about CCA and PSDU transmission status. It maintains the internal state of the "Network Allocation Vector" (NAV), which represents the availability of the radio media. It implements requests of control frames received from the radio media. It read the received PSDUs from the radio media via the PHY-SAP. It ignores the PDSUs that are not addressed to this station. It provides the PDSUs and data management to the defragmentation block.

3.1.6.5.1 Tx frame transmission sub-block

Tx frame transmission receives the MPDUs from the Fragmentation block and schedules them for transmission.
It provides signals to the sub-block "Sequencer" to indicate that a MPDU is ready for transmission. It accepts requests from the sub-block "Sequencer" to build RTS, CTS and ACK control frames.
The Tx frame transmission sub-block delivers PSDUs as soon as they are built, to the PHY, via the PHY-SAP.

112

It receives the PHY transmission status reports through the SAP-PHY.
It provides reports about transmission status for the MPDU processed, to the Fragmentation block.

3.1.6.5.2 Tx CRC Check
It eliminates the PSDUs with an incorrect CRC.

3.1.6.5.3 Tx duration extraction
It examines all PSDUs from the radio media that are addressed to this station and extract time information from these PSDUs. Time information is used to update the NAV value.

3.1.6.5.4 Tx Address Matching
It receives the PSDUs from the 'Duration Extractor' sub-block of the radio media. It provides PSDUs, that were sent to this station), to the Rx Ctrl / Data & Mgmt Demux sub-block. PSDUs that do not meet the criteria for comparison are not processed and are eliminated.

3.1.6.5.5 Tx Rx Ctrl / data & Mgmt Demux sub-block
It receives the PSDUs from the "Address Matching" sub-block; It provides the MSDU to the defragmentation block; It provides control frame to "Control decode" sub-block.

3.1.6.5.6 Tx control Decode
It receives control frames from the sub-block "Rx Ctrl / Data & Mgmt Demux". It generates control signals for the block "Tx Control / Sequencer" based on the contents of these control frames.

3.1.6.5.7 Sequencer
It receives signals from sub-blocks "Tx frame transmission", NAV and "Control Decode". It aggregates this information and generates signals to sub-block "Tx frame transmission" to initiate the transmission of data and control frames as part of a sequence of related frames.

3.1.6.5.8 Tx NAV

It maintains a correct Network Allocation Vector (NAV) state, based on information derived from the CCA information about the radio media and durations lengths derived from the block Duration Extractor".

It provides information to the Tx control / sequencer block to help in the planning of PSDU transmission on radio media.

3.1.6.6 PHY SAP

The PHY SAP is the Service Access Point (SAP) between MAC and PHY. PSDUs that must be transmitted by the station and PSDUs that are received by the station go through the SAP. The PHY also provides through the SAP signals for the MAC to indicate the status of "Clear Channel Assessment".

3.1.6.7 Defragmentation

It restores a full MSDU from fragments (MPDUs). It delivers the whole MSDU to the "Rx MSDU / MMSDU Demux block.

3.1.6.8 Rx MSDU Demux

It receives whole MSDUs from the defragmentation block; It passes an MSDU to the Rx Gating block.

3.1.6.9 Rx Gating

It accepts an MSDU received from the radio media, through the block "Rx MSDU Demux". It stores the MSDU which meet a number of criteria; MSDUs meeting the criteria are passed to the upper layers via the MAC-SAP; The MSDU that does not meet the criteria is eliminated.

3.1.7 DATA TRANSFER BETWEEN LLC AND MAC LAYERS

The MAC-SAP is the interface (Service Access Point) between the 802.11 MAC and upper layers in the protocol stack. An MSDU transmitted or received by the station goes through the MAC-SAP. The 802.11 defines three primitives for making possible this transfer through the MAC-SAP: MA-UNITDATA.request, MA-UNITDATA.indication and MA-UNITDATA.confirm.

There are used in a similar way, the management interface uses the MLME-SAP. The upper layers may request transmission of an MSDU by the primitive "MA-UNITDATA.request". The MAC layer generates a primitive "MA-UNITDATA.confirm" corresponding after transmission (or failure of transmission) of an MSDU. When the station receives an MSDU from the radio media, the MAC layer generates a primitive "MA-UNITDATA.indication".

3.1.7.1 MA-UNITDATA.request

This primitive transfers an MSDU, from a local LLC to the remote LLC, or groups of LLC.
Upon receipt of this primitive, the MAC layer determines if it is able to fulfill the request in accordance with the required parameters.
If the application can be completed in accordance with the required parameters, the MAC layer adds the specified fields, transmits the frame properly formatted to the PHY layer for the transfer to a peer MAC layer and indicates this action to the LLC sub layer by using a MA-UNITDATA-STATUS.indication primitive with the transmission status set to "Successful."
A request that cannot be completed in accordance with the required parameters is ignored, and this action is indicated to the LLC sub layer by using a MA-UNITDATA-STATUS.indication primitive that explains why the MAC layer could not meet the demand.

3.1.7.2 MA-UNITDATA.indication

This primitive transfers an MSDU from the MAC layer to the LLC layer entity or entities (in the case of the group addresses).

- In the absence of error, the content of the given parameter is unchanged from the parameter data of the primitive MA-UNITDATA.request.
- The reception status parameter indicates the success or failure of the frame received for 802.11 via a primitive frame MA-UNITDATA.indication. This MAC still reported "success" because all reception failures are ignored without generating the MA-UNITDATA.indication primitive.

3.1.7.3 MA-UNITDATA-STATUS.indication

This primitive has local significance and provides the LLC sub layer with status for the primitive information corresponding previous MA-UNITDATA.request

3.1.8 Instructions exchanged between MAC and PHY layers:

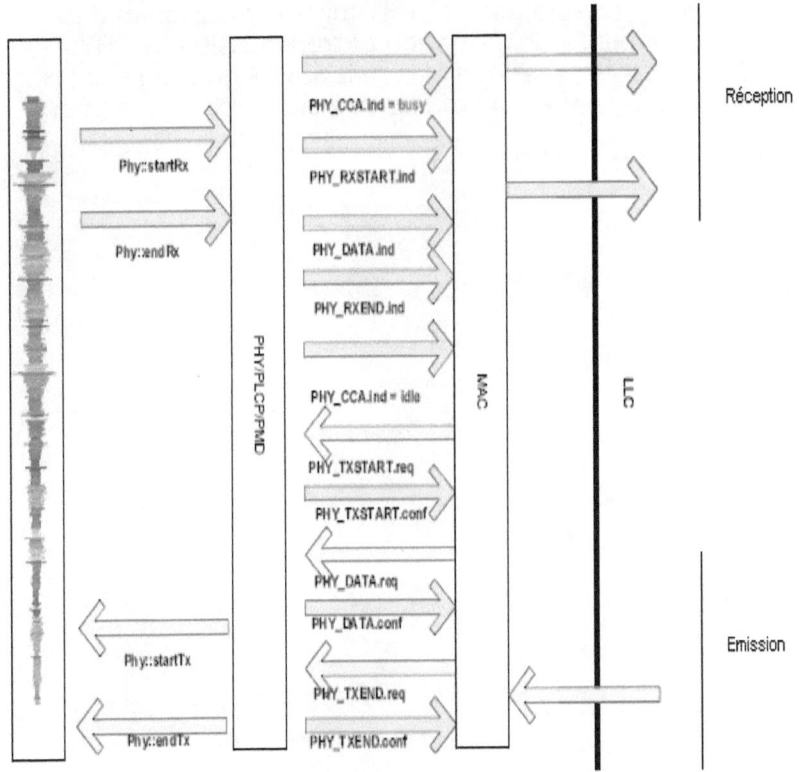

Figure 33 Primitives through 802.11 protocol stack

Reception:

PHY_CCA.IND - PHY layer sends the Clear Channel status (free / busy).

PHY_RXSTART.IND - Indicates that PHY has begun to receive; includes length and RSSI parameters.

PHY_DATA.IND - PHY indicates that the data arrives.

PHY_TXEND.IND - Indication by PHY that transmission is complete.

Transmission:

PHY_TXSTART.req - Tells the PHY to start the transmission; includes: length, data rate, service, TXPWR_LEVEL parameters.

PHY_TXSTART.conf - Confirmation of PHY that transmission began.

PHY_DATA.req - Request to PHY for transmission of data.

PHY_DATA.conf - Confirmation of transmission by PHY.

PHY_TXEND.req - "end of transmission" signal sent to PHY.

PHY_TXEND.conf - "end of transmission" confirmation of PHY.

3.2 MEDIA SHARING

Access to the radio medium is difficult because it can be subject to multiple reflections, echoes and Doppler effects. The collision detection was not possible in legacy 802.11 amendments because it was not possible to both listen and emit at the same time. The channel is shared between all stations but as a given station does not have access to a global view of the exchanges, some stations may be hidden from other. It is therefore not possible to use the collision detection Protocol which is used on Ethernet: CSMA/CD.

3.2.1 CLASSICAL IEEE.11 METHODS FOR SHARING MEDIA

The MAC layer defines two different access methods to combat the collisions:
• The method CSMA/CA used in the "Distributed Coordination Function" (DCF)
• The "Point Coordination Function" (PCF). PCF is an overlay above DCF. Unfortunately it was not implemented in most 802.11 chips.

3.2.1.1 CSMA/CA or DCF.

The 802.11 standard offers a protocol tailored for radio media, which is called CSMA/CA (Carrier Sense Multiple Access with Collision Avoidance).
As in radio it is not possible to transmit and receive at the same time, a station cannot detect a collision. So a procedure must be created for other stations to understand that they must stay away from the media when a transmission is ongoing. To that effect a receiver, if it received a correct frame must indicate this by sending an acknowledge frame to the issuer. While this is used as an error detection mechanism if missing, it is also used at the heart of CSMA/CA to forbid other stations to try to transmit as long as the ACK is not sent. Furthermore it shall not transmit before a time which is chosen at random, is elapsed. As different stations will probably choose different time durations, the odd that two stations are in contention is low. However this is disputed for number of stations above 6 because of the "birthday odds" that tells that in a given group, the odds that two members share a numerical characteristic is higher that intuition would suggest.
Nevertheless this CSMA/CA mechanism is useful in high latency

layers such as was 802.11b, because it forbids other stations to assume that a channel is free if no radio transmission occurs.

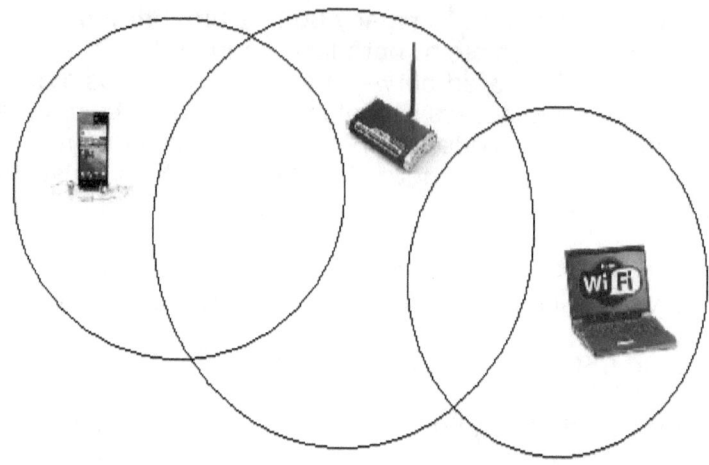

Figure 34 BSS and CSMA/CA

Here is a simplified view of the working of CSMA/CA.
(1) A station wishing to transmit, listens to the network. If the network is used, then the transmission is deferred after the ACK.
(2) Otherwise, if the media is still free after a given time (called DIFS for Distributed Inter frame Space), the station calculates a random number. If the channel is free for this random number multiplied by slot time then the channel is assumed to be free.
(3) The station starts sending data.
(4) Upon receipt of all data from the emitting station, the receiver sends an acknowledgment (ACK).
(1) All neighboring stations waiting for data transmission can start now at step 1 at their turn.

3.2.1.2 Point Coordinated Function

Point Coordinated Function is a technique for accessing the radio media that is used in the networks where a central station can organize the access to the channel. It is used only in infrastructure mode where an access point (AP), can indicate stations when they have the right to communicate. It is a bit strange that PCF is never implemented as it seems logical that the AP which is already a mandatory proxy between two stations that communicates with each other for confidentiality reasons, would also control and regulate the stations access to the media. As the access points and stations are connected on a single media, only a single message can be transmitted at the same time, however in MU-MIMO it is possible for several stations to communicate with the AP at the same time.

"CF Poll frame":
PCF uses a control frame system (CF Poll frame) to regulate stations access to the media. The Station Coordinator, actually the access point, sends this frame (CF Poll frame) in turn to each of the stations, to grant them the permission to transmit a single frame. If the queried station has no frame to send, it nevertheless must return an empty data frame with no data, called null data frame. If the station does not receive an acknowledgment frame, it cannot transmit the frame until the access point authorizes it. The Station Coordinator scans all the stations so they can access the media fairly. PCF has a higher priority than that of Distributed Coordination Function (DCF). A specific mechanism allows the stations that do not implement PCF to nevertheless inter operate on a BSS that implements PCF. PCF seems to be implemented only on few hardware devices as it is not part of the standard of interoperability of Wi-Fi Alliance. TXOP is a similar functionality that is perhaps more flexible in real traffic, while still being under the access point control.

3.2.2 BACKOFF PROCEDURE
3.2.2.1 classical Backoff
In older amendments a new transmission could only start after the DIFS (or RIFS) time plus a random time calculated by multiplying a random number by a contention window value (the slot time) which is often of 9 micro-seconds duration.

This time can be reduced with the RIFS.

3.2.2.2 Backoff in EDCA
It is an extension of the CSMA/CA "backoff" mechanism and it is one of the main features of EDCA. It is designed to solve access to radio media between several stations that try to access the media at the same time. EDCA uses a reduced DIFS, named AIFS, and each QoS class has its own AIFS duration. So some QoS classes enjoy priority over others. EDCA has also a backoff mechanism with different parameters depending on the class. When EDCA is used, each EDCAF (EDCA function) maintains its own variable contention window (CW) and delay timer.

3.2.3 THE CSMA/CA METHOD WITH CTS/RTS

DCF includes an option called RTS/CTS. RTS/CTS is an option of CSMA/CA (Carrier Sense Multiple Access / Collision Avoidance) that is used to ensure that a radio media is free and ready to be used, before transmitting a frame.

The RTS/CTS mechanism is a mechanism that a station uses to request the right to transmit on the media. It is therefore a collision avoidance mechanism based on a principle of acknowledgment by the access point:

The station wishing to transmit listens to the network. If the network is congested, then the transmission is deferred. Otherwise, if the media is free during a given time (called DIFS for Distributed Inter frame Space), the station can emit. The station transmits a message called Ready To Send (RTS) containing information about the volume of data it wishes to issue and its transmission speed. The access point sends a Clear To Send (CTS), and then the station starts sending data. Upon receipt of all data from the station, the receiver sends an acknowledgment (ACK).

Any station that "sees" an RTS will refrain from trying to transmit on the media as long as it will not be free again and a DIFS will be not passed.

However it is possible that at the time where a station is transmitting a RTS, another station issues another RTS, as the media seems to itself, to be inactive. But in this case the either stations will not receive any CTS because the access point has received nothing meaningful.

In order to minimize the probability of a collision of frames when two or more stations attempt to transmit at the same time, RTS is sent only when the data frame to be transmitted, is greater than a certain size. The RTS frame is actually sacrificed to test the availability of the channel (instead of sending a much larger block of data as in CSMA/CA). It's better to sacrifice a small frame such as RTS than a much longer data frame. This mechanism also helps mitigate problems such as "hidden stations" radio 802.11 networks.

In case of a multi-user transmission a BlockAck, is sent by each receiving station.

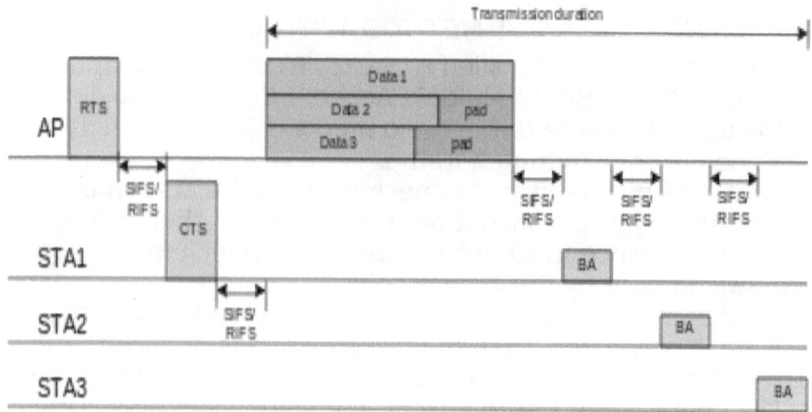

Figure 35 Multi-user transmission with a BlockAck

3.2.2.2 RTC/CTS Frame and Dynamic Channel Width Operation

RTS/CTS mechanisms were modified to improve the performance of the dynamic adjustment of the width of the channel. Consider a scenario of interference where the STA2 station transmitting at the access point AP2 and the AP1 communicates with STA1 and STA1 is in STA2 sight, but access points are not visible from the other.

AP1 <- some distance -> **STA1**<- some shorter distance ->**STA2** <- some other distance -> **AP2**

Let say the access point AP1 uses two channels, P being the primary channel and S1 the secondary channel and the same for AP2, except that the primary channel of AP2 is the secondary channel of AP1.

One channel	An adjacent channel	
P	S1	from the point of view of AP1
S1	P	from the point of view of AP2

Figure 36 Overlapping BSS (1)

AP2 access point main channel, is the secondary channel of AP1. A station must always use the same main channel than its access point. STA1 and STA2 stations can interfere with each

124

other, but the two access points being too far away, are not able to interfere with each other, nor to identify that the other access point and station are interfering with their station. To remedy to this situation, RTS and CTS frames are added to the usual control frames. AP1 sends a RTS with channel width and the various channels that make it up. Before STA1 responds with a CTS frame, it tries to detect if there are any radio activity on the secondary channel. If the secondary channel is not free, STA1 sends a CTS response with the only channel it perceives as free.

So the new situation is:

One channel	An adjacent channel	
P	S1	from the point of view of AP1
P	XX	from the point of view of STA1
S1	P	from the point of view of AP2

Figure 37 Overlapping BSS (2)

Then AP1 sends data to STA1 only on the available channel and STA1 responds with BlockAck (BA) frames which are duplicated on the free channels.

3.2.2.3 Operating Mode Notification Frame

In previous paragraph we saw how RTS/CTS could be used to tackle the problem of interference from another BSS.

However another new mechanism can be used. In this case, STA1 can send a frame "operating mode notification" to mean to the access point that the station wants to change the width of the channel in which it wishes to operate. For example, the STA1 station can change its width from 80 MHz to 40 MHz channel, with the restriction that the station must always use the same main channel as the access point. Subsequently, the access point will send only the data frames in this reduced channel width.

- Operating Mode Notification frame tells AP that the STA is changing the BW on which it operates

125

- ○ E.g. 80 MHz -> 40 MHz

- AP will then only send data frames occupying the reduced channel bandwidth

- Operating Mode Notification frame can also be used to reduce the number of spatial streams that a STA can receive (this is an enhancement of 11n's SM power save mechanism).

3.2.3 DISTRIBUTED INTER FRAME SPACE (DIFS)

Inter space frames or IFS (Inter frame Spacing), corresponds to an interval of time between the issuance of two frames. There are three types according to 802.11:

- SIFS (Short IFS), is the minimum time between the last symbol of a frame and the beginning of the first symbol of next frame. It is used mainly to reserve time between the transmission of an acknowledgment frame and the next frame to make it possible for the station that initiated the exchange that produced this ACK, to send further data frames if needed without other stations gaining access to the channel.
- DIFS (DCF IFS), which is used in DCF (I.e. in CSMA/CA) when a station wants to initiate a communication.
- PIFS, used by the access point to perform the polling in the PCF method.

There used to be a fourth one: RIFS which was the time between the last symbol of a frame and the beginning of the first symbol of next frame. RIFS is deprecated in 802.11ac.

3.2.4 NAV

The station internal information about the channel availability, is called NAV (Network Allocation Vector). It's also the name of a field in the duration field in the MAC frame control header. It is used to avoid collisions by delaying emissions of all stations even if they detect that the media is free.

It is used for the "hidden station" type situations where a station detects that the media is free while actually it is occupied by another station out of reach of this station radio. The access point can impose on all stations to not occupy the media, even if it looks to them to be free. It does so by setting the NAV indicator within the duration field in the MAC frame control header. Stations that do not see the hidden station, yet see the NAV information and they know so that the access point communicates with a hidden station. Even if it's very useful, NAV does not however solve all situations.

The NAV bit is also used in other cases such as in the PCF protocol.

It is also used when the CSMA/CA Protocol is extended with RTS/CTS signaling. All stations of the network receiving RTS or CTS frames will trigger their internal NAV indicator for a period corresponding to the time indicated in the duration field of the control header of the frame, to delay any transmission that they would have done. The transmitting station can then transmit and receive its acknowledgment without any risk of collision.

An 802.11ac station updates its NAV information by using the duration value of any frame in PPDU in the primary channel. A station does not update its NAV in response to a secondary frame, even if it is able to receive these frames.

3.2.5 QUALITY OF SERVICE (QoS)

802.11e enhances DCF and the PCF, through a new coordination function: the coordination function (HCF) hybrid. In the HCF, there are two methods for accessing the channel, similar to those defined in 802.11 MAC:

- Enhanced Channel Access (Distributed EDCA).
- HCF of the channel of access (HCCA) control

The two methods EDCA and HCCA define traffic categories (TC). For example, emails can be assigned a low priority class, and Voice over Wireless LAN (VoWLAN) could be attributed to a high priority class.

3.2.5.1 HCF/EDCA

With EDCA, high priority traffic is more likely to be sent to the recipient than low priority traffic: a station with traffic high priority waits a little less before gaining access to the channel before it sends its packet. This is accomplished by using a shorter contention window (CW) and a shorter arbitration time for high priority packets.

The levels of priority in EDCA are called access categories (ACS). There are eight traffic classes, four categories of access per station. It is a system based on queues in the station. Exact values depend on the physical layer that allows transmitting the data.

In addition, EDCA provides to a specific station, an unrestrained access to the channel for a period called "Opportunity Transmit" (TXOP).

TXOP is developed further in this chapter.

3.2.5.2 HCF/HCCA

HCF allows more complex prioritization schemes than in EDCA. EDCA does not predict when a station will have access to the media, as it indicates only when it's possible to transmit the information.

HCCA it is based on a centralized control by the access point which can guarantee the time and duration of transmission of information for each station. Each station which wants to associate with the access point must negotiate with it the level of QoS. The access point is then able to tell the station if there are sufficient resources in the BSS to meet its demand.

During a "Controlled Access Phase" the access point alone, manages the access to the media by the BSS stations. A station wishing to transmit information on the radio media, sends a «Reservation Request" to the access point. If the access point agrees, it informs the station of the time allocated.

So it's quite similar to TXOP.

The HCF (hybrid coordination function) of HCCA works somewhat like PCF.

However, unlike PCF, in which the interval between two beacon frames is divided into two periods between the CFP and CP, there is no allocated HCF/HCCA time slot. The HCCA allows PCF to be initiated almost in a CP. A CFP is initiated by the access point when it wants to send a frame to a station or receive a frame from a station without contention. In the CFP, the hybrid Coordinator (HC) - which is in the access point - controls the stations access to the media. In the CP, all stations operate in EDCA.

The other difference with the PCF, is that the traffic class (TC) and the traffic stream (TS) are defined. This means that the HC is not limited to a queue by station and can provide a service session queue. In addition, the CH can coordinate these streams or sessions in the way that he prefers (and not only in round-robin). In addition, the stations can give more information about the lengths of queues for each traffic class (TC). The HC can use this information to give more priority to a station than to another, or better adjust its scheduling mechanism. In the CP, the HC allows stations to send data by sending CF-Poll frames. HCCA is not required for access points. The implementation of HCCA on the stations uses the DCF mechanism for channel access (it is not necessary to change the operation mode of DCF or EDCA).

3.2.5.3 Service Classes.

In the QoS mode, a class of service for frames can have two values: QosAck and QosNoAck. Frames with QosNoAck are not acknowledged with an ACK frame. This avoids retransmission of time critical data.

Queues AC_xx and AAC_xx:

The 'priority' parameter for the data (payload) delivered on these queues, is specified in the MA-UNITDATA.request.

QoS supports eight priority values, referred to as UPs. One can take the '0' to 7 integer values and are identical to the IEEE 802.1 D priority tags.

The information is provided with each MSDU to the MAC_SAP either directly in the parameter to the top, or indirectly, in a TSPEC designated by the parameter to the top.

3.2.6 FRAME STRUCTURE:

There are three main types of frames:
- Frames for data used for the transmission of data
- Frames for control, for example RTS, CTS and ACK
- Frames for management, for the exchange of information management at the MAC level (association, authentication, and encryption).

General frame format

The maximum size is of the MPDU is given by the value of dot11FragmentationThreshold. It specifies the current maximum size, in octets, of the MPDU. an MSDU is fragmented when the resulting frame is larger than this threshold
The maximum size of an MPDU never exceeds 11500 or the aPSDUMaxLength of the PHY including the header and the FSC. MAC frames generally have the following format, which is possibly complemented by other fields:

Control header	Body	FCS

Figure 38 Main MAC frame structure

Each frame consists of a MAC header, the payload and frame check sequence (FCS). Some frames may not have payload. The first two bytes of the MAC header form a control field specifying the form and function of the frame.

3.2.6.1 Header control field

The header control has the following format and it contains key information on the frame.

frame control	Duration ID	Address 1	Address 2	Address 3	Sequence control	Address 4	QOS control	HT control
2	2	6	6	6	2	6	2	4

Figure 39 Header control

3.2.6.1.1 *Field of control of frame (frame control)*

The frame control field is probably one of the most important parts of the MAC. It's there that are managed all frame's feature.

Management	Association request
	Association response
	Re-association request
	Re-association response
	Probe request
	Probe response
	Timing Advertisement
	Reserved
	Beacon
	ATIM
	Disassociation
	Authentication
	Deauthentication
	Action
	Action No Ack
	Reserved
Control	11 Reserved
	beam-forming Report Poll
	VHT NDP Announcement

	Reserved
	Control Wrapper
	Block Ack Request (BlockAckReq)
	Block Ack (BlockAck)
	PS-Poll
	RTS
	CTS
	ACK
	CF-End
	CF-End CF-Ack
Data	Data
	Data CF-Ack
	Data CF-Poll
	Data CF-Ack CF-Poll
	Null
	CF-Ack (no data)
	CF-Poll (no data)
	CF-Ack CF-Poll (no data)
	QoS Data
	QoS Data CF-Ack
	QoS Data CF-Poll
	QoS Data CF-Ack CF-Poll
	QoS Null (no data)

	Reserved
	QoS CF-Poll (no data)
	QoS CF-Ack CF-Poll (no data)
Reserved	Reserved

Figure 40 Different types of frames

The frame control field is subdivided as follows:

Protocol version: these two bits represent the version of the Protocol. The version of the Protocol currently used is equal to zero. Other values are reserved for future use.

Type and subtype: these bits identify the 802.11 frame type (control, data and management) as well as the sub type (association, RTS, data, QoS). In 802.11ac there are two new under types: "VHT NDP Announcement" and "beam-forming Report Poll control frames.

ToDS and FromDS: these bits indicate whether a data frame is handled by a distribution system. Control and management frames set these values to zero. All data frames will have one of these bits to "one". However in an IBSS network these bits are always set to zero. When both bits are "one", it is the bridge function of the "wireless distribution system (WDS)" which is used in Mesh networks.

ToDS	FromDS	Add 1	Add 2	Add 3	Add 4
0	0	DA	SA	BSSID	
0	1	DA	BSSID	SA	
1	0	BSSID	SA	DA	
1	1	station with address RA	station with address TA	DA	SA

Figure 41 ToDS and FromDS sub-fields

More Fragments: The More Fragments bit is set to 'one' when a package for transmission is distributed over multiple frames. Each frame, except the last frame of a package will have this bit set to "one".

Retry: Sometimes it is necessary to retransmit frames. To achieve this, in a frame, often of control, there is a bit named 'Retry' which is set to "one" when one frame is retransmitted. As retransmitting frames may create situations where a frame is duplicated, this helps the frame receiver to identify duplicated frames.

Power management: this bit indicates the status of the station transmitter power management after the completion of a frame exchange with the access point. Access points use this information to manage the caching of frames as a station in power save can't receive any frame. Access points never have the right to "lift" the energy saving bit as it would be useless to send frames to a station in power save mode.

More data: the bit "More data" is used to point out that there are other frames cached in the access point and therefore in need to be transmitted. The access point uses this bit to facilitate the transition of stations in power saving mode to active state. The field "More data" is set to "one" in individually addressed frames transmitted by an access point to an 802.11ac station when both support the TXOP energy management feature to indicate that at least one frame is present in the buffer for the station.

Security (Protected frame field): the 'security' bit is enabled to "one" for an encrypted frame.

Order: This bit is normally set when the "strict policy" method of delivery of frames is used. Frames and fragments are not always sent in the normal order as it can sometime penalize the transmission performance.

3.2.6.1.2 Duration/ID:
The next two bytes are reserved for either the ID field or duration in microseconds. This field can take one of three forms: duration, period without contention (CFP), and the identity Association (AID). Today what it most often contains the value of the parameter NAV which is a period during which stations must refrain from starting a frame exchange because the access point favors an exchange with a specific station.

(a) In the PS-Poll subtype control frames, the duration/ID field contains the identifier association (AID) of the station which transmits a frame, in the 14 least significant bits (LSB). And the 2 most significant bits (MSB) of the duration/ID field are both set to "one". The value of the AID is in the range 1-2007.

(b) in frames transmitted by stations without QoS, in the CFP, the duration/ID field is set to a fixed value of 32768 (all bits at "one".

(c) in all other frames sent by stations without QoS and control frames sent by stations with QoS, the duration/ID field contains a duration value in micro seconds.

(d) for the data and management frames sent by stations with QoS, the duration/ID field contains a duration value.

As we will see have seen, it's important to aggregate MPDUs in a larger one because otherwise the new high throughput amendments would be very inefficient at PHY level. The duration/ID field of MPDUs transported in a single A-MPDU, itself inserted in one MU VHT PPDU, and all have the same value.

3.2.6.1.3 EDCA and simple and multiple protections:

In a frame transmitted under EDCA by a station that initiates a TXOP, there are two categories of parameters length: simple and multiple protections.

Simple protection: The frame duration/ID field value is used for the NAV duration value for protecting transmitting stations up to the end of a single frame. NAV is duration while it is forbidden for stations to access the media even if it is free.

Multiple protections: The value of the duration/ID of the frame field can define a time duration that protects up to the estimated end of a sequence of multiple frames. The frames that have the RDG/More PPDU sub-field at "one", always use multiple protections. The PSMP frames still use multiple protections. The station selects between single and multiple protection when it passes the first frame of a TXOP. All subsequent frames transmitted by the station in the same TXOP use the same duration settings class.

NDP announcement frames and report on sounding for beam forming frame always use multiple protection.

3.2.6.1.4) single and group Identity

The duration/ID field is determined as follows:

- (a) Single protection settings.
  - (1) For an RTS frame the duration/ID field is the time to transmit the current frame, plus one CTS frame, plus ACK frame or a BlockAck, if necessary, plus applicable IFS times.
  - (2) For all CTS frames the duration/ID field is set to one of the following:
    - (I) if there is a response frame to a previous NDP (sounding frame for beam-forming), it is set to the estimated time required to transmit the current frame, plus a SIFS time, plus the frame response (ACK or BlockAck), plus any NDP that were required, plus explicit feedback , plus an additional SIFs interval
    - (II) if there is no response frame, it's the time required to transmit the current frame, as well as a SIFs interval
  - (3) For a BlockAckReq frame, the duration/ID field is set to the estimated time required to transmit an ACK or BlockAck frame, depending on the case, plus a SIFs interval.
  - (4) For a BlockAck frame of, which is not sent in response to a BlockAckReq or an implicit Block Ack request, the duration/ID field is set to the duration estimated to transmit an ACK frame, plus a SIFs interval.
  - (5) For the following frames:
    - management frames
    - data frames without QoS (that is, with bit 7 of the Frame control field equal to 0)
    - data frames with the Ack policy sub-field equal to Normal Ack only,

The duration/ID field is set to one of the following:

- (I) if the frame is the last fragment of the TXOP, the duration/ID field is set to the estimated time necessary for the transmission of an Ack frame (including the appropriate IFS values)
- (II) in the other cases, the duration/ID field is set to the time necessary for the transmission of an Ack frame, as well as the time required for the transmission of the following MPDU, as well as IFS times.

- (6) For the following frames:
  - individual the data frames with the Ack policy sub-field equal to No Ack or Block Ack
  - the management frames with subtype Action set to No Ack
  - group addressed frame,

the duration/ID field is set to one of the following:

- (I) if the frame is the last fragment of the TXOP, it is set to "zero"
- (II) otherwise, the estimated time required for transmission of the next frame and its response frame if necessary (including the appropriate IFS values)

- (b) Multiple protection settings. The duration/ID field has a value as follows:
  - (1) If the station has no valid TXOP ongoing (it lost an unconditional access to the media) and it is neither forbidden to access the radio media, then the duration/ID sub-field is set to a value equal to the time required for a full MAC frame exchange.
  - (2) Else if the station has no valid TXOP ongoing (it lost an unconditional access to the media) but it is forbidden to access the radio media, then the duration/ID sub-field is set to a value equal to the time until the end of the NAV duration (as set by the access point).

141

- (3) Else it is not forbidden to access the radio media, then the duration/ID sub-field is set to a value equal to the time until the end of the TXOP duration
- (4) Otherwise the duration/ID sub-field is set to a value between NAV and TXOP.

3.2.6.1.5 Fields with address

A MAC 802.11 frame can have up to four address fields. Each field can carry a MAC address. A MAC address is a sequence of 6 bytes (often represented as hexadecimal example: 01: 23: 45: 67: 89: AE) that uniquely identifies each network interface. The first address is often those of the receiver, the second address is usually that of the transmitter, the third address is usually used for filtering purposes by the receiver.

A MAC address has one of the following two types:
- 1. Individual address. The address assigned to a specific station on the network.
- 2 Group address. An address to multiple destinations, which can be used by one or more stations on a given network. The two types of group addresses are as follows:
  - Multicast group address. An address bound to a group of logically related stations.
  - Broadcast address. A group address which always matches all stations in a given network. All zeros are interpreted to be the broadcast address. It is used to broadcast to all active stations on this radio media.

TA Field address station contains an individual MAC address which identifies the station which transmitted the MPDU.

If individual/ group bit is 0, the address TA docking station field indicates an individual address.

otherwise, the TA field address identifies a sounding station indicating that the frame contains additional information in the coding sequence.

Group ID works as follow:

    10 AP transmits MU MIMO PPDU to a group of STAs identified by Group ID

    20 STAs use Group ID to index local table to identify its Nsts (# of SS) Index

    30 Nsts (# of SS) Index determines which space-time streams the STA demodulates

3.2.6.1.6 Sequence Control field

The sequence control field is a two byte field that is used to identify the order of the messages as well as to enable the elimination of duplicate frames. The first 4 bits are used for to indicate the number of fragments of an MSDU into MPDUs and the last 12 bits are the sequence number of a MPDU. It is similar to a page numbering scheme such as "page 58 over 251".

3.2.6.1.7 QoS field

Two optional bytes for QoS have been added with 802.11e. One is the "QoS" field, the other field the '" HT/VHT control" field. The QoS Control field is present in all data frames in which the QoS sub-field of the Subtype field is equal to "one".

Quality of service control field is a 16 bit field that identifies the TC or TS to which belongs the frame as well as various other QoS information on the frame which varies by frame type, subtype and the type of transmitting station. The field of control of the quality of service is present in all frames of data in which the QoS of the subtype field sub-field is equal to one. Each quality of service control field consists of five or eight sub domains.

The sub-field of Ack policy is 2 bits in length and it indicates whether or not an acknowledgment is necessary in the received frame.

- a) in a frame that is a non-A-MPDU frame or a single 802.11ac MPDU:

143

- The addressed recipient returns an Ack frame after a SIFS.
- For a station this is the only value allowed for the policy sub-field for acknowledgment for the individual QoS frames.
- (b) in other cases (especially for a frame that is part of an A-MPDU), the recipient returns a BlockAck MPDU, either individually or as part of an A-MPDU starting from SIFS after the PPDU carrying the frame.

3.2.6.1.8 HT/VHT Control field

To exploit the multi-rate capability of 802.11ac, a wireless station should possess the capability to dynamically select the best transmission rate based on the network condition and channel quality. A basic knowledge that a station needs is the signal to noise ratio of a spatial stream. A more complex situation arises when a station aims at creating a beam toward another station or to know the paths properties of spatial streams, sends or request a more sophisticated mechanism to the station at the other side of the path. That station will send/request a sounding PPDU, which is a PPDU without any data (NDP). The characteristics of a PPDU without any data are perfectly known by the receiver station, so the alterations this station found, indicates how the spatial stream alters the emitted signal.

However the NDP mechanism is very slow, a NDP transmission should occur for each spatial stream. Sometimes it's preferred to extend the basic mechanism described above to multi-spatial streams, where an estimation of the SNR and MCS is done for each spatial stream. The field «HT/VHT control" deals with all those situations. The field «HT/VHT control" is the last of two optional bytes to the QoS that were added with 802.11e. The other is the "QoS" field.

The HT/VHT control field is always present within a 32-bit group. The first bit indicates whether it is a "HT control" field belonging to a VHT frame or not. 31 next bits are the classic field "HT control". Actually it is the bit zero of the former "Link Adaptation Control" field that is used to do this.

A MAC frame contains a HT (high throughput) control field, which further contains:

* A MCS request (MRQ) sub-field,

* an MCS sequence identifier (MSI) sub-field,

* an MFB sequence identifier/LSB of Group ID (MFSI/GID-L) sub-field,

* a VHT N_STS, MCS, BW and SNR feedback (MFB) sub-field,

* an MSB of Group ID (GID-H) sub-field, coding type of MFB response (Coding Type) sub-field, transmission type of MFB response (FB Tx Type) sub-field, unsolicited MFB sub-field, AC constraint sub-field, and RDG/More PPDU sub-field.

The MFB sub-field further contains a number of spatial streams (N_STS) sub-field, an MCS sub-field, a bandwidth (BW) sub-field, and an SNR sub-field.

- Solicited MFB: When the MRQ sub-field is set to one, MFB is then requested, and the MSI sub-field contains a sequence number in the range from zero to six that identifies the specific request.

- If the unsolicited MFB sub-field is set to one, then MFSI/GID-L sub-field contains the lowest three bits of Group ID of the PPDU to which the unsolicited MFB refers, and the GID-H sub-field contains the highest three bits of Group ID of the PPDU to which the unsolicited MFB refers. When MFB is requested, the MFB sub-field contains the number of spatial streams, modulation and coding scheme, data transmission rate, bandwidth, and SNR information.

The receiving device provides the MFB sub-field feedback information for every valid sub-channel of an 802.11ac channel. When a sounding is used for the 802.11ac channel, the MFB sub-field contains multiple MFB sub-fields, and each MFB sub-field corresponds to a valid sub-channel. Each MFB sub-field contains N_STS/MCS/BW/SNR information for the corresponding sub-channel. For the MCS sub-field, the IEEE 802.11 standard defines several modulation and coding schemes with different data transmission rate. In one example, estimated SNR is used to perform an effective link adaptation based on the network condition and channel quality. Link adaptation based on SNR is a PHY-aware MAC implementation that allows the MAC layer to select a PHY data rate based on estimated SNR and desired packet error rate. The MCS sub-field comprises modulation type, channel-coding type, channel-coding rate, spatial rank, and transmission diversity type.

Field HT (802.11n)

	B0 B15	B16 B17	B18 B19	B20 B21	B22 B23	B24	B25 B29	B30	B31
	Link Adaptation Control	Calibration Position	Calibration Sequence	Reserved	CSI/ Steering	NDP Announcement	Reserved	AC Constraint	RDG/ More PPDU
Bits:	16	2	2	2	2	1	5	1	1

Figure 42 Field Link Adaptation Control in 802.11n

Field VHT (802.11ac)

Figure 43 Field Link Adaptation Control in 802.11ac

	B1	B2	B3 B5	B6 B8	B9 B23	B24 B26	B27	B28	B29
	Reserved	MRQ	MSI/ STBC	MFSI/ GID-L	MFB	GID-H	Coding Type	FB Tx Type	Unsolicited MFB
Bits: 1		1	3	3	15	3	1	1	1

In VHT, most of the HT indications are not used:

- TRQ: Not Needed (VHT supports explicit FB only)

- Calibration Position: Not Needed (VHT supports explicit FB only)

- Calibration Sequence: Not Needed (VHT supports explicit FB only)

- CSI/Steering: Not Needed

- NDPA: Not Needed (VHT supports explicit NDPA Frame)

New fields are defined:

- **MRQ**: Request for feedback about MCS. MRQ is used for VHT link adaptation.

The value is set to "1" when requesting a VHT-MCS (solicited MFB) report, otherwise the value is set to "0".

- **MSI/STBC**: MRQ sequence identifying / STBC indication
A few words about STBC: Space–time block coding is a technique used in wireless communications to transmit multiple copies of a data stream across a number of antennas and to exploit the various received versions of the data to improve the reliability of data-transfer. The fact that the transmitted signal must traverse a inhomogeneous environment means that some of the received copies of the data will be 'better' than others. This redundancy results in a higher chance of being able to use one or more of the received copies to correctly decode the received signal. Space–time coding attempts at using all the copies of the received signal in an optimal way to extract as much information from each of them as possible. STBC has fewer options in 802.11ac than in 802.11n.

- If the Unsolicited MFB sub-field is 0 and the MRQ sub-field is 0, the MSI/STBC sub-field is reserved.

- If the Unsolicited MFB sub-field is 0 and the MRQ sub-field is 1, the MSI/STBC sub-field contains a sequence number in the range 0 to 6 that identifies the specific MCS feedback request.

- If the Unsolicited MFB sub-field is 1 and the MFB does not contain the value representing "no feedback is present",
  - The MSI/STBC field contains the Compressed MSI and STBC Indication sub-fields.

  - The STBC Indication sub-field indicates whether or not the estimate in the MFB sub-field is computed based on a PPDU using STBC encoding:

    - Set to 0 if the PPDU was not STBC encoded

    - Set to 1 if the PPDU was STBC encoded

  - The Compressed MSI contains a sequence number that identifies the specific MCS feedback request. It is in the range 0 to 3 if STBC Indication equals 0 or in the range 0 to 2 if STBC Indication equals
- Otherwise, the MSI/STBC sub-field is reserved.

- **MFSI/GID-L**: MFB sequence identifier/LSBS of group ID

  - If the unsolicited sub-field MFB is 0,
    - the sub-field MFSI/GID-L contains the value received from MSI in the frame to which refers the MFB information.

148

- If the unsolicited sub-field MFB is '1' and the MFB does not contain the value that represents "no feedback is present" and the MFB is estimated from a MU PPDU VHT,
    - If the unsolicited MFB is estimated from a PPDU SU, the sub-field MFSI/GID-L is set to all zeros.
    - Otherwise, the sub-field MFSI/GID-L contains the 3-bit lower the group ID of the PPDU where the MFB was estimated.

- Otherwise, this sub-field is reserved.

- **MFB**: NUM_STS, VHT-MCS, BW and SNR feedback
  The MFB sub-field is interpreted as:
    - Recommended NUM_STS.
It indicates the recommended NUM_STS. The NUM_STS sub-field contains an unsigned integer representing the number of space time streams minus 1.
    - Recommended VHT-MCS.
It indicates the recommended VHT-MCS. The VHT-MCS sub-field contains an unsigned integer in the range 0 to 9 representing a VHT-MCS Index value.
    - BW.
It is the bandwidth of the recommended VHT-MCS

If the Unsolicited MFB sub-field is 1, the BW sub-field indicates the bandwidth for which the recommended VHT-MCS is intended:
- Set to 0 for 20 MHz
- Set to 1 for 40 MHz
- Set to 2 for 80 MHz
- Set to 3 for 160 MHz and 80+80 MHz.

If the Unsolicited MFB sub-field is 0, the BW sub-field is reserved.

- SNR:

It is the average SNR. It indicates the average SNR, which is an SNR averaged over data sub-carriers and space-time streams. The SNR is averaged over all the space-time streams and data sub-carriers, and is encoded as a 6-bit two's complement number of SNR_average. This encoding covers the SNR range from –10 dB to 53 dB in 1 dB steps.

The MFB sub-field contains the recommended MFB. The combination of VHT-MCS = 15 and NUM_STS = 7 indicates that no feedback is present.

- **GID-H**: group ID MSBs

If the Unsolicited MFB sub-field is 1 and the MFB does not contain the value representing "no feedback is present" and the unsolicited MFB is estimated from a VHT MU PPDU, the GID-H sub-field contains the highest 3 bits of group ID of the PPDU from which the unsolicited MFB was estimated (bit 3 of the group ID appears in the lowest numbered bit of the field GID-H). If the unsolicited MFB is estimated from an SU PPDU, the GID-H sub-field is set to all 1s.
Otherwise this sub-field is reserved.

- **Coding Type**: Code type of measured PPDU

If the unsolicited sub-field MFB is '1' and the MFB does not contain the value that represents "no feedback is present", the encoding Type sub-field contains the information coding ('0' for BCC) and '1' for LDPC of PPDU from which the unsolicited MFB was estimated.
Otherwise this sub-field is reserved.

- **FB Tx Type**: Type of Transmission of measured PPDU
  - If the unsolicited sub-field MFB is '1':
    - If the MFB does not contain the value representing "no feedback is present" and sub-field FB Tx Type is '0',

150

- the unsolicited MFB is estimated from a PPDU VHT with RXVECTOR's BEAM-FORMED parameter equal to 0.

- If the MFB does not contain the value that represents "no feedback is present" and FB Tx Type sub-field is 1,
  - The unsolicited MFB is estimated from a PPDU VHT with the RXVECTOR BEAM-FORMED parameter equal to '1'.

- Otherwise this sub-field is reserved.

- **unsolicited MFB**:

Unsolicited feedback VHT-MCS indicator
When the MRQ sub-field is set to one, MFB is then requested. The value "1" indicates that the MFB is not a response to an MRQ.
Otherwise the value is set to '0'.

3.2.6.2 Frame body

The frame body field is of variable size, and it contains information from higher layers.

The classic size body of frame (pre-802.11n) is based on the maximum size an MSDU: 2304 bytes. The minimum size is zero. In 802.11n, the maximum size of the body frame is determined by an MSDU (2304 bytes) maximum size or the maximum size of the MSDU (3839 or 7935 bytes, depending on the station capacity).

In 802.11ac, 4 692 480 octets is the maximum length in octets for a VHT SU PPDU with a bandwidth of 160 MHz or 80+80 MHz, VHT-MCS9 and 8 spatial streams, while the size of an A-MSDU may be variable.
A-MSDU length = length max MPDU (114.5) - length of header MAC (36) - overhead Protocol CCMP (16) - FCS (4) = 11 398 bytes.

	Non-HT station PPDU and non-HT duplicate PPDU	HT PPDU	VHT PPDU	DMG PPDU
MSDU size	2304	2304	2304	7920
A-MSDU size	3839 or 4065	3839 or 7935		7935
MPDU size			3895 or 7991 or 11 4.5	
PSDU size	$(2^{12})-1$	$(2^{16})-1$	4 692 480	$(2^{18})-1$
PPDU duration		54.5	54.5	2000

Figure 44 MPDU size

If a management MPDU is sent using an 802.11ac PPDU, the length of the MPDU is limited by the maximum length supported by the PPDU

3.2.6.3 Check Sequence

The frame Check Sequence (FCS) consists of the last four bytes of the 802.11 MAC frame. Often referred to as the "Cyclic Redundancy Check (CRC), it allows to verify the integrity of retrieved frames. The FCS is calculated and added when the frames are about to be sent. When a station receives a frame, it can calculate the FCS of the frame and compare it to what it really received. If they match, it is assumed that the frame has not been corrupted during transmission.

3.2.7 CONTROL FRAMES

There are a dozen control frame, the best-known is the ACK frame (used to report that the frame was well received), the RTS and the CTS frames (used to avoid collisions). The control frame format is special; each frame has its own, often short format.

3.2.7.1 Acknowledging good reception (ACK)

After receiving a data frame, if no error is detected, the receiving station sends an ACK frame to the transmitting station. If the sending station does not receive an ACK frame within a predetermined period, it sends again the data frame.

Frame control	Duration	Receiver Address	FCS

Figure 45 ACK frame

3.2.7.2 Request to send (RTS)

The RTS and CTS frames provide an optional collision detection system. A station sends a RTS frame and waits a CTS frame before sending the data frame.

The RTS frame is intentionally of very small duration, so if it is crushed by another station, there is only a little penalty on the additional imparted delay to the station with respect of what it would have been with a collision during a data frame. The other stations that see a RTS on the radio media (stations constantly scan the radio media) should wait for trying to access it. They should wait until the ACK frame, then the additional inter-frame delay plus the contention window time.

If the RTS is mainly a query to obtain access to the radio media, and is intended mainly at the other stations, CTS it mainly intended to the RTS emitter. It tells it that the access point correctly received the RTS, and grants it the right to use the radio media. Therefore a station receiving a CTS, knows it now owns the radio media for a frame duration.

RTS/CTS/ACK scheme is not used every time, first because it is not useful is some situations like "hidden stations", but also because it is a bit verbose. This was even worse with the use of aggregated channels because of the need to send RTS/CTS/ACK on every channel.

When an 802.11ac device sends an RTS, this device has to verify that the aggregated channel is clear in its vicinity, the RTS is normally sent in an 802.11a PPDU format, and the basic *802.11a* transmission, which is 20 MHz wide, is replicated to fill all the aggregated channels. Then every nearby station, regardless of whether it is an 802.11a/n/ac station, receives an RTS that the station can understand on its primary channel. RTS/CTS/ACK competition is TXOP/NAV which are tokens for monopoly/exclusion from the radio media. But those concepts are intermingled for example; using RTS and CTS frames provides the benefit of auto reset of the Network Allocation Vector (NAV). For example, if a receiver receives a RTS but does not receive a corresponding CTS or Data frame for the RTS, the receiver will reset its NAV.

Frame control	Duration	Receiver Address	Transmitter Address	FCS

Figure 46 RTS frame

3.2.7.3 RTS with bandwidth indication

When an 802.11ac device sends an RTS, this device has to verify that the aggregated channel is clear in its vicinity, the RTS is normally sent in an 802.11a PPDU format, and the basic *802.11a* transmission, which is 20 MHz wide, is replicated to fill all the aggregated channels. Then every nearby station, regardless of whether it is an 802.11a/n/ac station, receives an RTS that the station can understand on its primary channel. To ensure that all stations hear and respect the Duration field included in the RTS and CTS frames, the frames should be backwards compatible with legacy protocols. For example, if sending RTS/CTS frame on a 5 GHz channel, a frame should be sent that is compatible with 802.11a. The PLCP header of the a

RTS frame or a CTS frame comprises a rate field, a reserved bit, a Duration field, a parity bit, and tail bits for the preamble of the PLCP header. The RTS/CTS frame described here is compatible with legacy 802.11a RTS and/or CTS frames.

When sending a duplicated Request To Send (RTS) frames across 40/80/160 MHz channels, Clear To Send (CTS) frames should be returned for the same channel. But some channels may be busy. So the transmitter may decide to send the PPDU only on the available channels and it had to indicate which channels are used. This is the bandwidth indication. RTS is a nice place to set this information as RTS is sent on each 20 MHz channel and not on the full aggregated channels.

When a receiver receives the packet, the receiver can detect whether the frame contains a bandwidth parameter set. For example, the receiver checks the status of the multicast/unicast address of the TA field. If the TA field contains data indicating the packet contains bandwidth parameter set data, the receiver runs a scrambler in reverse to obtained the partitioned scrambler seed, and obtains the bandwidth indication, e.g., 20/40/80/160 MHz from the partitioned scrambler seed. In some cases, the receiver also determines whether the indication is static or dynamic.

3.2.7.4 Clear to Send (CTS)
A station or an access point responds to an RTS frame with a CTS frame. This frame provides permission for the requesting station that allows it to send a frame of data on the media. The CTS frame includes a value of time during which the stations non-involved in the exchange must not make attempts to access the radio media.

Frame control	Duration	Receiver Address	FCS

Figure 47 CTS frame

3.2.7.5 NDPA

NDPA is a new command frame to announce a request for assessment of the radio media characteristics to a station at the next frame. This assessment will be done with a Null Data packet, which will enable the receiver station to calculate the differences between the perfect and real characteristics of the spatial stream under study. In turn, when reported to the sender station, those differences will enable the sender station to take appropriate measures to compensate the imperfections in the radio media in the spatial stream.

Announcement of NDP VHT raster format

Frame
Control Duration RA TA
Sounding Dialog Token
station Info 1
...
station Info *n*
FCS

Figure 48 NDP Announcement frame

The "NDP Announcement" frame contains at least a field Info station. If the "NDP Announcement" frame contains only a single field Info station, the RA field contains the address of the identified station. If the "NDP Announcement" frame contains more than one field Info station, then the RA field contains the broadcast address.

3.2.8 MANAGEMENT FRAMES,

They are used for the exchange of information of MAC-level management. The management frame format is the same as the data frame format. Management frames allow the establishment and maintenance of the possibilities of communication within the cell:

Authentication frame:
A sequence of 802.11 authentication begins by sending an authentication frame containing the station's identity, at the access point. In the case of an open authentication the station sends only an authentication frame and the access point responds to an authentication frame by its acceptance or rejection. With shared key authentication, after the station has sent its initial authentication request, it will receive from the access point, an authentication frame with the challenge. The station sends a frame containing the encrypted version of the challenge to the access point. The access point verifies that the text was encrypted with the correct key, by trying to decipher it with its own key. The result of this process determines the status of the station authentication.

Association request frame: It is sent from a station to the access point. The frame carries information about the station including the supported rates and the SSID of the network to which the station wishes to associate. If the application is accepted, the access point reserves the memory and establishes an Association identifier for the station.

Association response frame: It is sent from an access point to a station and contains the acceptance or rejection of an application for association. If there is an acceptance, the frame will contain information about the association such as the MCS.

Beacon frame: Sent periodically from an access point to announce its presence at the stations just arrived in the vicinity and provide the SSID, and other parameters such as time synchronization to stations within its reach.

Those parameters are:

- 1 Time-stamp
- 2 Beacon interval
- 3 Capability
- 4 Service Set Identifier (SSID)
- 5 Supported rates
- 6 Frequency-Hopping (FH) Parameter Set
- 7 DSSS Parameter Set
- 8 CF Parameter Set The CF Parameter Set element is present only within beacon frames generated by APs supporting a PCF.
- 9 IBSS Parameter Set The IBSS Parameter Set element is present only within Beacon frames generated by stations in an IBSS.
- 10 Traffic indication The TIM element is present only within Beacon frames generated map (TIM) by access points or mesh stations.
- 11 Country
- 12 FH Parameters
- 13 FH Pattern Table
- 14 Power Constraint
- 15 Channel Switch
- 16 Quiet
- 17 IBSS
- 18 TPC Report
- 19 ERP
- 20 Extended Rates Supported
- 21 RSN
- 22 BSS Load
- 23 EDCA Parameter
- 24 QoS Capability
- 25 AP Channel Report
- 26 BSS Average Access
- 27 Antenna

- 28 BSS Available
- 29 BSS AC Access
- 30 Measurement Pilot
- 31 Multiple BSSID
- 32 RM Enabled
- 33 Mobility Domain
- 34 DSE registered
- 35 Extended Channel
- 36 Supported Operating
- 37 HT Capabilities
- 38 HT Operation
- 39 20/40 BSS Coexistence
- 40 Overlapping BSS
- 41 Extended Capabilities
- 42 FMS Descriptor
- 43 QoS Traffic
- 44 Time Advertisement
- 45 Interworking
- 46 Advertisement
- 47 Roaming
- 48 Emergency
- 49 Mesh ID
- 50 Mesh Configuration
- 51 Mesh Awake
- 52 Beacon Timing
- 53 MCCAOP Advertisement Overview
- 54 HT Capabilities MCCAOP One or more MCCAOP Advertisement elements
- 55 Mesh Channel Switch Parameters
- ...
- 60 VHT Capabilities
- 61 VHT Operation
- 62 VHT Transmit Power One
- 63 Channel Switch

- 64 Extended BSS Load
- 65 Quiet Channel
- 66 Operating Mode

Figure 49 Beacon frame parameters

Of-authentication frame: Sent from a station to an access point because it wants to put an end to its authentication and therefore its session within the cell.

De-association frame: Sent from a station wishing to terminate its session within the BSS. It is an elegant way to allow an access point to delete the memory allocation for this station and remove the association table.

Information request frame: Sent from a station when it requires information about an access point.

Information response frame: Sent by an access point, after receiving a request from the station frame, it contains information of capacity, the maximum data transmission rate supported, etc.

Re-association request frame: A station sends a re-association request when it is out of range of the access point currently associated and that he found an another access point with a signal stronger. The new access point coordinates transmission of any information that could be contained in the buffer of the previous access point.

Re-association response frame: Sent by an access point, it contains the acceptance or rejection of a station re-association request frame. The frame includes information such as the Association identifier and supported rates.

3.3 MANAGEMENT FRAME

3.3.1 WHY AGGREGATING FRAMES

At PHY and MAC-level there are a number of timers with fixed
durations that had very reasonable values in 802.11b but which
are very inadequate with the vast throughput increase. For
example the duration of a PHY preamble in 802.11b/g is
192μsec, however as the throughput shrinks more and more
the information part of the frame in the meantime, this constant
size of the preamble defeats the throughput increase. The
following table illustrates the progressive deterioration in the
effectiveness of the MAC layer.

	Preamble	information
802.11b	1%	99%
802.11g	8%	92%
802.11n 20Mhz MIMO 2*2	29%	71%
802.11n 40Mhz MIMO 2*2	45%	55%
802.11n 40Mhz MIMO 4*4	73%	27%

Figure 50 Relative growing share of preamble with higher
throughput amendments.

In newer technologies the PHY frame which has a constant
length, is less and less used to convey information. Most of the
PPDU is empty. The reason why we keep these preamble
lengths is that this is essential to allow interoperability between
different technologies. By recognizing that interoperability
doesn't mandate MAC level interoperability, it's possible to
aggregate several MAC frames in a legacy physical frame
(PPDU). So for example multiple 802.11ac MPDU could be
aggregated in on a single 802.11a MPDU and a single PPDU.

If you send multiple MPDU in one PPDU frame, the risk of
collisions, the time lost in Backoff timeouts or in management
frames is also reduced.

Of course a limitation of the frame aggregation mechanism is that all frames that are grouped into a transmission should be sent to the same destination.
Another limitation is that all frames to aggregate must be ready at the same time, which could delay some frames, in an attempt to send a single overall frame.

The maximum frame size is increased in 802.11n, to accommodate the aggregated frames. The maximum frame size is increased from 4 KB to 64 KB. In 802.11ac it's as large as one megabyte.

3.3.2 A-MSDU AND A-MPDU

Improving the throughput at the physical level also involves reducing the verbosity of the protocol. Frame aggregation is a mechanism that can improve the performance. There are two types of frame aggregation.

Frame aggregation is a process of aggregation of multiple MSDU frames into an A-MSDU frame or a process of aggregation of multiple MPDU which will generate an A-MPDU frame. This in order to reduce the management dialogue and thereby increasing the data rate.

In practice 802.11n or 802.11ac stations often implement only A-MPDU. However with the rise of throughput of 802.11ac, the implementation of these two functions has become necessary to offset the harmful effects of incompressible times such as DIFS. However only one error is enough to cancel the transmission of multiple frames and thus to considerably reduce the throughput.

Because the address information is in the MAC header, the PHY layer (which considers it as a part of its payload) processes all the frames that it is able to detect on the media. It belongs to the MAC layer to determine if the receiving station is concerned with the decoded frame or not.

3.3.2.1 MAC Service Data Unit Aggregation (A-MSDU)

A MSDU often carries 1500 bytes. Logical groupings of MSDU packets with the same 802.11e quality of service are done at MAC level, independently of the source or the destination. The newly built MAC frame then contains a MAC header, followed by a maximum of 7935 bytes of MSDUs in 802.11n. In 802.11ac the size of A-MSDU passes to 11454 bytes. The groupings are done at the top of the MAC layer. The resulting A-MSDU is encapsulated in an MPDU.

From a station, the aggregated frame is sent at the access point, where the aggregated MAC frames are dis-assembled and transmitted to their respective destination.
From an access point, all frames included in the aggregated frame must be sent for a single station.

The aggregation of an MSDU is often the most effective of the two aggregation methods. It is based on the fact that MSDUs are "logical" units, whereas MPDUs are simply fragments of a MSDU. When emitting there is no meaning in fragmenting a MSDU in several MPDUs, then partly aggregating them in an A-MPDU.
Theoretically, frames with different destinations can be batched together in a single frame. In practice, however, an MSDU aggregation collects only the MAC frames with a common destination, encapsulates these different frames in a single PHY frame and forwards this associate PHY frame. This method is more efficient than the MPDU aggregation up to the permitted maximum of size a 3839 MSDU or 7935 bytes (depending on the capacity of the client), because the MAC header is much shorter than the PHY header.

3.3.2.2 MAC Protocol Data Unit Aggregation (A-MPDU)

An A-MPDU is a sequence of MPDU sub-frames encapsulated in a single PPDU.

MAC Protocol Data Unit (A-MPDU) aggregation occurs after the creation of the A-MSDU.

The MPDU groupings are realized at the bottom of the MAC layer. The MSDUs are fragmented in the MPDU; MAC headers are added to each MSDU. The MPDUs are then aggregated into A-MPDU.

Figure 51 With and without MPDU aggregation

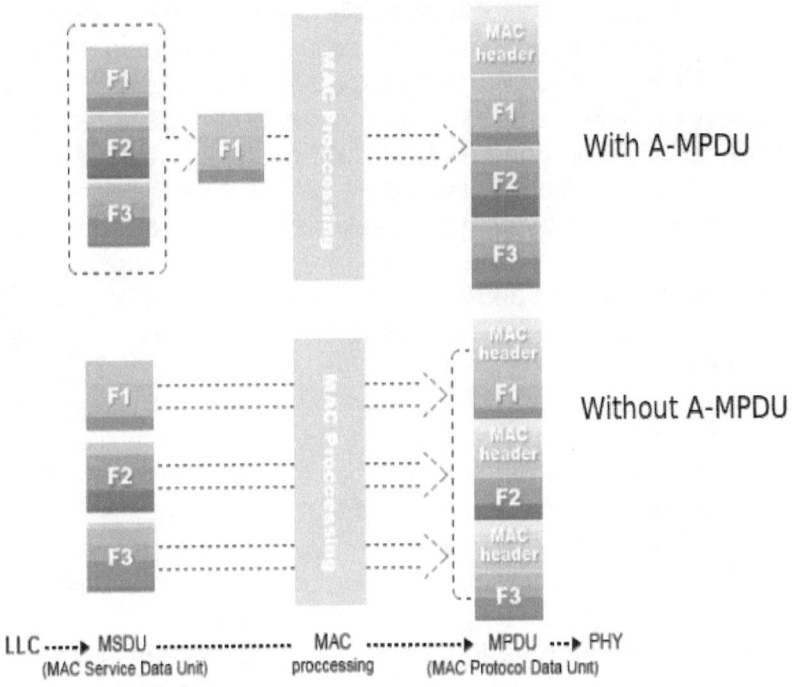

MPDU aggregation selects the frames that have a common destination. The A-MPDU frame is then integrated within a PHY frame that may have a size up to 65535 bytes in 802.11n, and a mega-byte in 802.11ac.

A-MPDU frames require the use of the "block acknowledge" that was introduced in 802.11e and has been optimized in 802.11n. These protocol extensions enable to increase the physical layer (PHY) data rate, but can't be used to communicate with other pre-802.11n stations. With pre-802.11n stations the classic ACK should be used to confirm an A-MSDU, but block acknowledge should be used to confirm reception of the aggregated MPDU. MPDU aggregation does not require that all frames that construct up the aggregated frame, should have the same destination address. However, this result in the same behavior as in MSDU aggregation, since the destination of all frames sent by the station are to the access point of this station, where frames are then forwarded to the final destination.

With MPDU aggregation, it is possible to encrypt each frame independently of others, using the security association SA for each destination MAC address. There is no effective difference in encryption with respect to MSDU aggregation, because all frames sent by a station are encrypted by using the access point security association, and all frames sent by the access point are encrypted by using the security association to the station that will receive the frame.

In a similar manner to an MSDU aggregation, the MPDU aggregation requires that all constituent frames have the same level of QoS.

The effectiveness of the MPDU aggregation method is less than the MSDU aggregation method for short and medium size clusters, due to the additional number of headers of each MAC frame embedded in the aggregated MPDU. Efficiency is further reduced if frames are encrypted. Encryption adds an overlay to each MPDU of the aggregated MPDU then an aggregated MSDU which is encrypted has only one of those overlays.

However, MPDU aggregation is the preferred schema when large amounts of data are available for the aggregation.

In 802.11ac all physical frames must be aggregated, even if there is only a single physical frame. This is necessary because the PHY layer does not any more contains the size but only the number of OFDM symbols, and one MPDU frame contains only the duration and not the size.

3.3.2.3 format A-MPDU

The format A-MPDU in 802.11ac is an extension of the 802.11n A-MPDU. A-MPDU maximum length in an 802.11ac PPDU is 1.048.576 bytes.

An 802.11ac station that transmits a VHT PPDU, contains one or more PSDUs, each of which contains an A-MPDU, for one or more stations.

168

An A-MPDU is composed for each user from any of the following:

— A-MPDU sub-frames constructed from the MPDUs available for transmission that have a TID value that maps to the primary AC

— A-MPDU sub-frames with 0 in the MPDU Length field and 0 in the EOF field

The A-MPDU_Length[n] for station n is initialized as the length of the resulting A-MPDU pre-EOF padding.

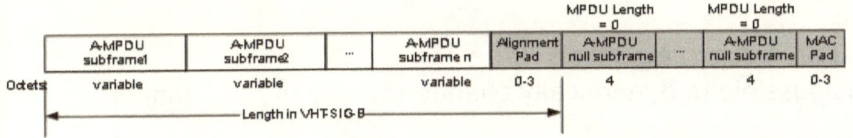

Figure 52 A-MPDU format for PPDU 802.11ac

An 802.11ac A-MPDU fills all the available bytes in the payload. The extension (in gray) consists of zero or more delimiters with zero length MPDU and final possible EOF of less than 4 bytes for alignment.

An A-MPDU sub-frame is composed of

- A MPDU delimiter

- A MPDU

- EOF

End of frame (EOF) bit indicates that there is no present additional MPDU in A-MPDU.

3.3.2.3.1 Format of A-MPDU delimiter
The A-MPDU delimiter is amended as indicated in figure 16.

B0	B1	B2 B3	B4 B15	B16 B23	B24 B31

EOF	Reserved	MPDU Length Extension	MPDU Length	CRC	Delimiter Signature

Figure 53 **Modified A-MPDU delimiter format**

An MPDU extension length field is added in B2 - B3 and contains the high bits of the MPDU length. An EOF field is added to B0.

3.3.2.4 Aggregation

It is possible to dynamically change the size of the channels.

Improvements in coexistence mechanisms exist to allow greater channel width in contexts mixing 802.11 a/n and 802.11ac 802.11ac essentially changes the 802.11n MAC layer management to address coexistence issues induced by the use of wider channels. In addition minor changes are made in the 802.11n mechanism of aggregation to improve efficiency. With 802.11n many channels at 20 and 40 MHz in the 5 GHz band, the risk of channels overlap between two BSS, can be easily avoided by choosing a different channel for each BSS. In the worst case, if an overlap is inevitable on the 40 MHz channel within two neighboring BSS, the same 20 MHz primary channel is chosen to allow a maximum of coexistence. With a much larger channel size in 802.11ac, it becomes much more difficult to avoid overlap between neighboring BSS. It also becomes more difficult to choose a primary channel common between overlapping BSS.

To resolve this issue, 802.11ac brings three improvements:
- better detection of occupation of the secondary channel with "Clear Channel Assessment (CCA),
- improving the management of dynamic width of the channel,
- a new frame of notification of the operation mode.

3.3.2.5 Block acknowledgment

This is to reduce the number of ACK frames that a station must send an issuer to confirm delivery of a frame. The old 802.11a/b/g stations expect an almost immediate ACK for every frame except those of multicast/broadcast. But an 802.11n station can also accept an ACK block that confirms the receipt of multiple frames. Reducing the number of ACKs that a receiver must send to a transmitter to confirm delivery of the frames help free bandwidth for really useful frames.

This feature is very useful for acknowledging multiple MPDUs aggregated in a A-MPDU.

Block acknowledgment is also used to acknowledge data sent over aggregated channels.

3.4 Linux MAC

3.4.1 Introduction

Most of WLAN card manufacturers follow the SoftMAC approach, much more flexible compared to the old FullMAC solution. FullMAC leaves all the control of the MAC layer functions to the card hardware/firmware, whereas SoftMAC implements a new set of control MAC primitives at the software level. The framework *mac80211*, which is part of the Linux 802.11 stack provides SoftMAC capabilities. Nevertheless not all physical hardware makes it possible to use a SoftMAC. In addition most SoftMAC capable hardware uses a binary file hiding implementations details to some degree. The situation is similar to video hardware were there are very few open source drivers. There are very few completely 802.11 open source drivers and as far I know, no 802.11ac open source drivers. This book will use a fictitious 802.11ac open source driver based loosely on Atheros hardware.

The mac80211 *was* adopted in Linux 802.11 networks to reduce the complexity and shorten the time when introducing new hardware devices.

The Linux 802.11 stack specifies the framework *mac80211* that enables SoftMAC-capable device drivers used for operating with 802.11 hardware. While some of the MAC functionalities are implemented in binaries or even at the hardware level, *mac80211* implements features such as handling several higher-layer components of the MAC, including support for HW/SW cryptography, power saving,.11n style aggregation. The mac80211 module plays two key roles:

- Wrap the packet incoming from the upper layers and translate them into the 802.11 frame format;

- Control management operations related to the IEEE 802.11 standard.

173

3.4.2 CFG80211

This layer exists between the user space and protocol driver (mac80211). These set of APIs perform sanity check and protocol translation to configure wireless devices. It provides functions for

- Device registration

- Regularity enforcement

- Station management

- Key management

- Mesh management

- Virtual Interface management

- Scanning

Device registration includes band, channel, bit rate, high throughput (HT) capabilities and supported interface modes. Regularity enforcement will ensure during the registration of cfg80211 that only the specified frequency channels permitted for that given country will be enabled. Station management include add, remove, modify stations and dump station details. These functions are part of AP capabilities.

In mesh path handling, mesh parameter set and retrieve are the functions provided for mesh management. Virtual interface management provides create, remove, change type and monitor flags. It also keeps track of the network wireless interface. Scanning allows user level initialization of scanning and reporting.

3.4.3 SOFTMAC TRANSMISSION PATH

SoftMAC supports IEEE 80211 a/b/d/g/n and s, different types of interfaces and QoS. The types of interfaces include STA, AP,

174

monitor and mesh. It handles the following protocol functionalities:

- Transmission Path

- Receive Path

The higher layer transfers the packet structure to the MAC by calling kernel public transmission function. The higher layer packet will be converted to IEEE 802.11 frame format and initializes all its required buffers and headers. The transmit handlers selects the key, transmission rate, inserts the sequence number (based on the hardware capability), selects the encryption algorithm, fragmentation, calculates the transmission time and generates control information for transmission.

3.4.4 SOFTMAC RECEIVE PATH

Hardware driver sends the frame to the protocol driver mac80211 along with the hardware receive status information.

In the receive path: mac80211 checks the type of the packet, receive status and prepares the receive handlers. Receive handler verifies the alignment of the packet for proper processing, decryption and defragmentation. The frames with aggregation, control frame (Block Acknowledgment), next management frame exchange are processed. The IEEE 802.11 frames are converted to IEEE 802.3 + LLC and sent to the higher layer.

3.4.5 SOFTMAC RATE ADAPTATION

Most of the current *mac80211* drivers rely on rate control algorithms provided by the framework. These algorithms are encapsulated in independent kernel modules that are linked to the specific driver once a new device is being loaded. *mac80211* implements two rate adaptation schemes, Minstrel and PID, but also permits the drivers to implement specific rate adaptation mechanisms and register them upon device initialization to notify the *mac80211* framework that rate selection will be handled by the driver itself.

Despite the differences of these two approaches, they both use a common mechanism to interface with the *mac80211* framework. Specifically, the *rate_control_ops* callbacks are registered by the rate adaptation module to the framework.

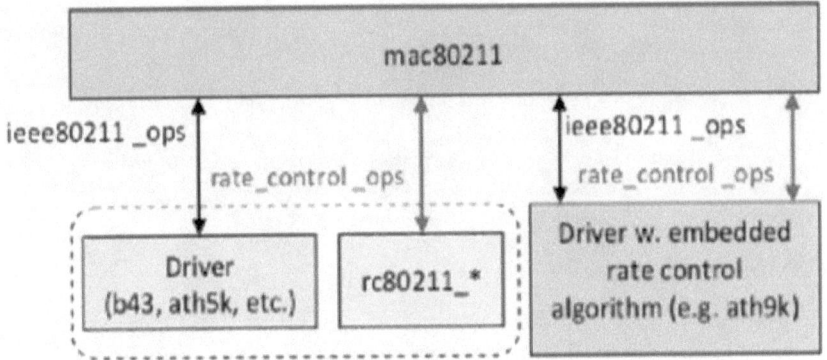

Figure 54: Rate control, MAC operation codes

4 PHY LAYER (PHY)

The physical layer is responsible for carrying, on the radio media, the information to be transmitted by the emitting station to the receiving station. The protocol that manages the physical layer (PHY) defines a method for transforming the MAC frames into PHY layer frames, allowing the transmission and reception of information on the radio interface.

The PHY layer provides three levels of functionality:

- The PHY layer provides a basic frame management service to the MAC layer under the control of the sub layer "physical layer convergence protocol" (PLCP). For example, it is able to manage PHY frames training fields with different throughput and understand the information in the SIG field. In recent versions of 802.11 (11n and following), this layer also manages channel aggregation, MIMO channel sounding and other MIMO aspects.
- In emission, the PHY layer codes and modulates a radio carrier to transmit data frames on the media under the control of the PMD sub layer. In reception, it synchronizes to the incoming signal and transforms of radio signals into digital data destined for the PLCP layer. However, in the standard 802.11ac it was argued that no separation should be made between PLCP and PMD.
- The PHY layer provides a carrier detect indication to the MAC layer to verify the activity on the radio media. Thus enabling CSMA/CA at MAC layer.

In transmission:

The packet information, obtained from the TXVECTOR, together with the MAC protocol data unit (PDU) are enough to generate the entire PHY packet.

- Regarding the data symbols, the first step is to create the Data field. In the next step, the bits from the data field are scrambled. The scrambler output bits are processed by the forward error correcting (FEC) encoder, which has two flavors: binary convolutional code (BCC) or low-density parity-check (LDPC) code.
- After this, the encoded bits are sent to the parser, which is responsible for the separation of the coded bits into spatial streams.
- In the next step, the bits are divided into symbols. The interleaving is applied to the coded bits for each OFDM symbol.
- In the following, the bits are mapped in constellation points, resulting in complex data symbols transported at each OFDM sub-carrier.
- Pilots are inserted and then each OFDM symbol is converted to time domain using the inverse Fast Fourier Transform (IFFT).
- Next, the CS is applied in each symbol at each transmitting antenna. The CP is added using the final samples of each OFDM symbol.
- The preamble is put in front of the data field, creating a complete VHT packet.

In reception:

- The first block is the packet detector, which uses the L-STF to detect the arriving of a packet.
- After packet detection, it is necessary to obtain the time synchronization, i.e., to estimate the position of the OFDM symbols inside the packet. The time synchronization is performed using the L-STF and L-LTF symbols. The frequency offset, estimated using these

same fields, is used to correct the remaining fields of the MF-HT packet. It is implemented a time domain auto-correlation algorithm to detect the packet, synchronize the OFDM symbols and to estimate the frequency offset.

- The next step is to perform the FFT of the L-LTF to estimate the channel. However, this procedure does not allow estimating a MIMO channel, therefore, this first channel estimation is used only to detect the L-SIG and VHT-SIG fields, which are single streams ones. The procedure to obtain the control information in the L-SIG and VHT-SIG symbols is the same used for the data symbols, except that the scrambler is not performed in the SIG fields.

- The procedure to receive the data symbols is as follows. First, the CP is removed, remaining only the "n" time samples of each OFDM symbol. These samples are used to perform the "n"-point FFT for each OFDM symbol.

- The MIMO channel estimation is implemented in the frequency domain using the VHT-LTFs. The sub-carriers are equalized using this MIMO channel estimation. The pilot symbols transmitted at each OFDM data symbol are used to estimate and correct the residual phase due to the imperfect time and frequency synchronization. The block called demapper performs hard or soft demodulation of the data symbols transport by the OFDM sub-carriers.

- The de-interleaving and de-parser reverse the corresponding operations performed at the transmitter side.

- Next, it operates the iterative decoding.

- Finally, the decoded bits go into the de-scrambler to recover the streams of transmitted bits.

4.1 MANAGEMENT FUNCTION OF THE PHY LAYER (PLME)

4.1.1 PHY LAYER MANAGEMENT ENTITY (PLME)

The PHY 802.11ac contains three functional entities: the PHY function, PLCP and PMD, and the management function of the physical layer (PLME). The PLME manages the local PHY functions in conjunction with the MLME. It provides to the MAC layer mostly the following protocol functions:

- A function that defines a mapping of the PSDUs in a framing (PPDU) format to send and receive PSDUs between two or more stations.
- A function that defines the characteristics and the mode of transmission and receiving data over a radio media between two stations. The PLME mostly deals with the PHY frame format.

The PHY provides an interface for the MAC layer through the extension of the generic PHY service interface. The PHY layer receives from the MAC layer the bits to be transmitted and the TXVECTOR, which contains control information about the packet (e.g., packet length, channel bandwidth, CP length, MCS). The interface enables transmission of information with TXVECTOR, RXVECTOR and PHYCONFIG_VECTOR as information containers. The structure of the PPDU transmitted by an 802.11ac station is determined by its TXVECTOR parameters. At the reception, the PPDU parameters received from the MAC layer in RXVECTOR inform the PHY layer.

With the help of PHYCONFIG_VECTOR, the MAC layer configures the PHY layer for the operation to be performed.

If TXVECTOR or RXVECTOR had a physical reality as names a long time ago, nowadays they are not visible as such in Linux code. More so the information and code is actually buried in binaries that are provided by manufacturers, often because FCC makes it mandatory for the user to not have too much control over the radio properties.

Nevertheless some information which is available in userland such as RSSI comes from the PHY.

An 802.11ac PPDU can be either a SU PPDU (single user) or MU PPDU (multi-user) depending on the value of the group ID. A MU 802.11ac PPDU can carry one or more PSDUs to one or more stations. It's common in the specification to designate stations in a multi-user session by the "user" term.

When transmitting, a PSDU (if single user) or one or more PSDUs (in the case of multi-user) are processed (i.e., scrambled and encoded) and added to the preamble to create the PPDU PHY. On the receiver, the PHY preamble is processed in order to assist in the detection, demodulation, and delivery of the PSDU.

4.1.2 *PHY-SAP* SERVICE

The PHY service is provided to the MAC layer through a SAP, called the PHY-SAP. A set of primitives might also be defined to describe the interface between the PLCP sub layer and the PMD sub layer, called the PMD_SAP.

PHY-DATA.request

This primitive defines the transfer of an octet of data from the MAC sub layer to the local PHY entity.

This primitive is generated by the MAC sub layer to transfer an octet of data to the PHY entity. This primitive can only be issued following a transmit initialization response (a PHY-TXSTART.confirm primitive) from the PHY.

The receipt of this primitive by the PHY entity causes the PLCP state machine to transmit an octet of data. When the PHY entity receives the octet, it issues a PHY-DATA.confirm primitive to the MAC sub layer.

PHY-DATA.indication

This primitive indicates the transfer of data from the PHY to the local MAC entity. The PHY-DATA.indication primitive is generated by a receiving PHY entity to transfer the received octet from the radio media to the local MAC entity.

PHY-DATA.confirm

This primitive is issued by the PHY to the MAC layer when the PLCP has completed the transfer of data from the MAC layer to the PHY. The PHY issues this primitive in response to every PHY-DATA.request primitive issued by the MAC sub layer. The receipt of this primitive by the MAC causes the MAC to start the next MAC layer request.

PHY-TXSTART.request

This primitive is a request by the MAC sub layer to the local PHY entity to start the transmission of a PSDU. The primitive uses the TXVECTOR parameter. This primitive is issued by the MAC sub layer to the PHY entity when the MAC sub layer needs to begin the transmission of a PSDU.

The effect of receipt of this primitive by the PHY entity is to start the local transmit state machine

PHY-TXSTART.confirm

This primitive is issued by the PHY to the local MAC layer to confirm the start of a transmission and to indicate parameters related to the start of the transmission. The PHY issues this primitive in response to every PHY-TXSTART.request primitive issued by the MAC sub layer.

This primitive uses TXSTATUS as parameter

The TXSTATUS represents a list of parameters that the local PHY entity provides to the MAC sub layer related to the transmission of an MPDU. This vector contains both PLCP and PHY operational parameters.

This primitive is issued by the PHY to the MAC layer once all of the following conditions are met:

— The PHY has received a PHY-TXSTART.request primitive from the MAC entity.

— The PHY is ready to begin accepting outgoing data octets from the MAC.

The receipt of this primitive by the MAC layer causes the MAC to start the transfer of data octets. Parameters in the TXSTATUS vector may be included in transmitted frames so that recipients on multiple channels can compensate for differences in the transmit time of the frames, and so to determine the time differences of air propagation times between transmitter and receiver and hence to compute the location of the transmitter.

186

In addition, the TXSTATUS vector may include the TX_START_OF_frame_OFFSET.

PHY-TXEND.request

This primitive is a request by the MAC sub layer to the local PHY entity that the current transmission of the PSDU be completed.

This primitive PHY-TXEND.request has no parameters.

This primitive is generated when the MAC sub layer has received the last PHY-DATA.confirm primitive from the local PHY entity for the PSDU currently being transferred. The effect of receipt of this primitive by the local PHY entity is to stop the transmission state machine.

PHY-TXEND.confirm

This primitive is issued by the PHY to the local MAC entity to confirm the completion of a transmission. The PHY issues this primitive in response to every PHY-TXEND.request primitive issued by the MAC sub layer.

The semantics of the primitive are as follows: "PHY-TXEND.confirm". This primitive has no parameters.

This primitive is issued by the PHY to the MAC entity when the PHY has received a PHY-TXEND.request primitive immediately after transmitting the end of the last bit of the last data octet indicating that the symbol containing the last data octet has been transferred and any Signal Extension has expired.

The receipt of this primitive by the MAC entity provides the time reference for the contention backoff protocol.

PHY-CCA.indication

The primitive provides the following parameters:

```
PHY-CCA.indication(

STATE,
```

```
IPI-REPORT,

channel-list

)
```

This primitive is generated within aCCATime of the occurrence of a change in the status of the channel(s) from idle to busy or vice-versa.

The STATE parameter can be one of two values: BUSY or IDLE. The parameter value is BUSY if the assessment of the channel(s) by the PHY determines that the channel(s) are not available. Otherwise, the value of the parameter is IDLE.

The IPI-REPORT parameter is present if dot11RadioMeasurementActivated is true and if IPI reporting has been turned on by the IPI-STATE parameter. The IPI-REPORT parameter provides a set of IPI values for a time interval. The set of IPI values may be used by the MAC sub layer for Radio Measurement purposes. The set of IPI values are recent values observed by the PHY entity since the generation of the most recent PHY-TXEND.confirm, PHY-RXEND.indication, PHY-CCARESET.confirm, or PHY_CCA.indication primitive, whichever occurred latest. When STATE is IDLE or when, for the type of PHY in operation, CCA is determined by a single channel, the channel-list parameter is absent. Otherwise, it carries a set indicating which channels are busy. The channel-list parameter in a PHY-CCA.indication primitive generated by an 802.11ac station contains at most a single element.

Primary channel

It indicates that the primary 20 MHz channel is busy (the electromagnetic activity on the channel is above the CCA sensitivity for signals occupying the primary 20 MHz channel).

For a 40 MHz channel, the primary channel is composed of this channel and the other 20 MHz channel, same schema is applicable for 80 and 160 MHz channels.

Secondary channel, or secondary channel at 40 MHz, or secondary channel at 80 MHz channel

It indicates that the secondary channel is busy (the electromagnetic activity on the channel is above the CCA sensitivity for signals not occupying the primary 20 MHz channel).

4.1.3 LINUX PHY

There is mostly not much code at this level as a lot is done in hardware. Nevertheless we will study the high level features and how they map to 802.11 specification.

A device driver for IEEE 802.11ac based wireless devices often uses mac80211 as the protocol driver.

Transmission Path

Since a packet is received at the driver from mac80211, the driver initializes the required buffers and maps it to the hardware queues. Transmit flags are assigned depending on the type of the packet, physical layer parameters and control information. The packets are transferred to the hardware through the interconnection driver using DMA. Transmit interrupt initiates for the transfer of the frame from the hardware and checks the status for reporting and retry.

Receive Path

When a packet is received by the hardware, it generates a receive interrupt. The driver function which is mapped to receive interrupt will generate required locks for fetching the packets from hardware and transferring it to protocol stack mac80211 with status information.

190

4.2 PHY LAYER PARAMETERS :

Transmission OFDM	
Frequency bands	5 GHz
Channel bandwidth (data sub-carrier)	M: 20 MHz (52) O: a 40 MHz (108)
OFDM symbol duration	M: 4 µs O: 3.6 µs (short GI)
Modulation	M: BPSK up to 64-QAM
FEC	M: BCC O: LDPC
Code rates	M: 1/2, 2/3, 3/4, 5/6
MIMO: Spatial Streams	M: 1, 2 (APs), direct correspondence O: 3, 4, TxBF, STBC
PHY Data rates	M: 6.5 –65 (APs: 130) Mb/s O: 6 –600 Mb/s
Spectral efficiency	0.3 –15 bit/s/Hz

Figure 55 PHY Timings

4.3 PLCP FUNCTIONS

4.3.1 PLCP IN EMISSION
When a PSDU comes from the MAC, it comes up with a set of parameters, which are collectively known as the TXVECTOR.

The TXVECTOR settings are determined by the MAC layer. It contains:

1. Information throughput
2. Length
3. Preamble type
4. Modulation
5. Transmission power level

One can see on the two figures below the various states through which passes the physical layer in emission and the messages exchanged between the physical layer and MAC layer.

- The PLCP asks the PMD to switch in transmission mode, after receiving the primitive PHY-TXSTART.request
- PLCP prepares a preamble and warned the MAC layer when it is finished. PHY-TXSTART.confirm is then sent.
- The MAC layer sends the data bytes (0-4095), and the throughput instructions, in a loop:
    - Loop_begin:
    - PHY-DATA.request
    - PHY-DATA.confirm
    - Test if end of loop.
        - If not then goto Loop_begin:
- PLCP adjusts the last elements of the preamble based on the data (size and checksum). The PLCP informs PMD.
- The PMD should respond by sending the frame preamble on the antenna in the next 20ms.
- The PMD emits the preamble and header, then the rest at the specified rate

- When it is finished, the PMD makes a report to PLCP which similarly reports to the MAC layer. This is done with PHY-TXEND.request and PHY-TXEND.confirm.

4.3.2 PHY TRANSMISSION PROCEDURE

There are two ways to transmit information at the PHY layer:
- The first way, which is typical for transmitting information at the PHY layer is selected if the FORMAT parameter TXVECTOR the primitive PHY-TXSTART.request (TXVECTOR) is VHT. The first way does not describe optional features such as LDPC, STBC or MU.
- The second way is to follow the 802.11a PHY transmission procedure if the FORMAT parameter of the communication primitive PHY-TXSTART.request (TXVECTOR) is NON_HT and NON_HT_MODULATION parameter is set NON_HT_DUP_OFDM. An alteration to this rule is that 802.11a PHY is rather generated at the same time on each of 20 MHz channels that are specified by parameter CH_BANDWIDTH as defined in the VHT preamble and Non-HT duplicate transmission.

Regarding the multi-user operation, there is an A-MPDU per user in the MAC sub layer, and the VHT training symbols plus VHT-SIG-B field plus and the DATA field are given per the user in the PHY layer. The number of VHT training symbol number depends on the total number of space-time flow for all users.

In both paths, to transmit data, the MAC layer generates a primitive PHY-TXSTART.request, which causes the switching of the PHY layer state machine in the transmitting state. In addition, the PHY is set to operate at the appropriate frequency via the PLME. Other transmission parameters, such as MCS types and transmission power are defined via the PHY-SAP using the PHY-TXSTART.request(TXVECTOR) primitive.

The PHY indicates the status of the primary channel and other channels (if applicable) via the primitive PHY-CCA.indication. PPDU transmission is initiated by the PHY after receiving the PHY-TXSTART.request (TXVECTOR) primitive.
After starting the transmission of the PHY preamble , the PHY layer immediately starts data transmission. The encoding method for the data field is based on the FEC_CODING, CH_BANDWIDTH, NUM_STS, STBC, MCS and NUM_USERS

parameters of the TXVECTOR. SERVICE field and PSDU are encoded.

The data will be exchanged between the MAC and PHY layer through a series of PHY-DATA.request (DATA) primitives issued by the MAC layer and PHY-DATA.confirm primitives issued by the PHY. The encoded PSDU is a multiple of the number of coded bits per OFDM symbol.

Transmission may be interrupted prematurely by the MAC layer through the primitive PHY-TXEND.request.

PSDU Transmission ends when receiving a PHY-TXEND.request primitive. Each PHY-TXEND.request correctly processed is followed with a primitive PHY-PHY TXEND.confirm. In single user transmission, normal termination takes place after the transmission of the last bit of the last byte PSDU as indicated by Nsym.

The PHY guard interval is inserted into each OFDM symbol. Once completed the PHY PPDU transmission switches to reception.

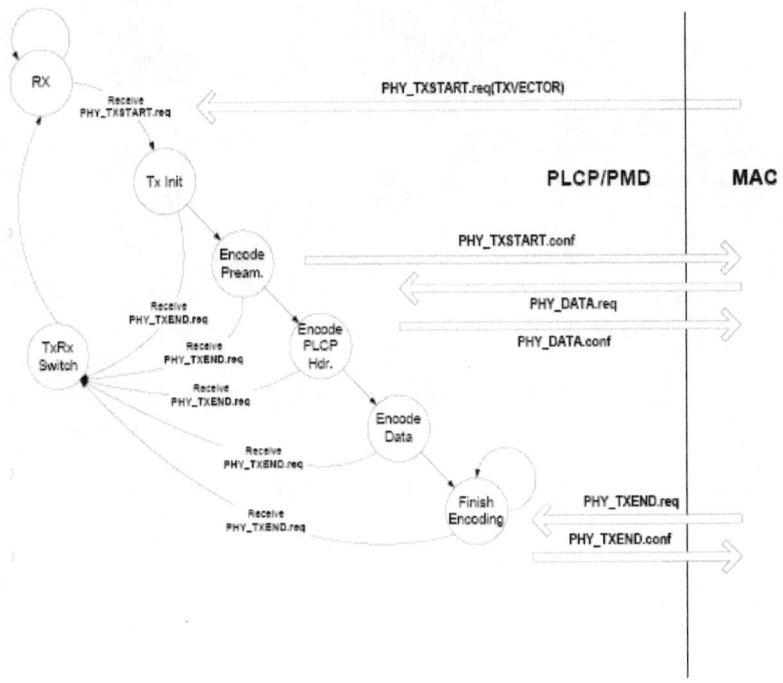

Figure 56 PHY/MAC interactions in transmission

4.3.3 PLCP IN RECEPTION

Receive mode is more complicated because frame reception is not scheduled in advance, the station must wait for any transmission, analyze it and send the received frame at the MAC layer.

When a PPDU is processed by PHY layer from the channel, the PHY layer extracted the PSDU from the PPDU, and sends it to the MAC layer with a set of parameters, which are collectively known as RXVECTOR.
The RXVECTOR settings include:
- Rate, duration, type of preamble, of modulation, of service,.
- Received Signal Strength Indicator (RSSI)
- At the end of the reception of the PPDU, it is also indicated in the MAC layer if the PPDU was correctly received or not:
- Reception error: violation of format, carrier lost;
- not supported throughput

If the medium is detected as being in busy mode, the PMD sub layer tries to decode a preamble.
If the preamble is error-free and that the clock has been recovered, PMD informs PLCP with a message PMD_RSSI.ind and PMD_DATA to indicate that the next transmission will focus on the PLCP preamble.
When the PLCP preamble was received by the PMD, it warns PLCP. PLCP will recover its preamble and inform the MAC with PHY_RXSTART layer
At the end of the transmission of the PSDU, the PHY layer sends PHY_TXEND to the MAC layer.

4.3.4 PHY RECEPTION PROCEDURE

The procedure described here to receive at PHY, does not describe machine operating states of optional features such as LDPC and STBC. It only applies to VHT PPDU format.

Through the PHY management interface (the VHT PLME) the appropriate frequency is set. The PHY is also configured with the group information (group membership and position in the group) so that it can receive data destined to the STA. Other receives parameters such as RSSI and data rate, can be accessed through the PHY-SAP interface.

When receiving the PHY preamble PHY measurement of signal strength reception on main channel. This activity is indicated by the PHY to the MAC via PHY-CCA.indication primitive. A primitive PHY-CCA.indication (BUSY, channel-list) is also published in a first estimation of the signal reception. The parameter of the channel list is present when the channel width is 80 MHz, 40 MHz, or 160 or 80 + 80 MHz.

The PHY includes the most recently measured RSSI value in PHY-RXSTART.indication (RXVECTOR) primitive sent to the Mac. After PHY-CCA.indication (BUSY, channel-list), the PHY layer will begin to receive training symbols and seeking L-SIG field to define the maximum data stream.

If the L-SIG field is not valid, a PHY error condition PHY-RXEND.indication (Format Violation) primitive is sent. If the L-SIG field is valid,, the PHY will maintain updated information for a CCA.indication (BUSY, channel-list) primitive for the expected duration of the transmitted PPDU, as defined by the RXTIME for all supported modes.

After receiving a valid L-SIG and VHT-SIG-A indicating a mode supported, the PHY layer begin to receive VHT training symbols and VHT-SIG-B

If the Group ID and partial AID fields in VHT-SIG-A has a value that indicates a multi-user VHT STA capable of been part in a beam forming. If the VHT-SIG-B indicates a mode not supported, the PHY delivers the primitive PHY-RX-END.indication (UnsupportedRate) error condition.

For single user VHT PPDU, the SU coding field / MU [0] VHT-SIG-A2 indicates the type of encoding. The PHY entity must use a LDPC decoder to decode C-PSDU for each user if this bit is set, otherwise the BCC decoding is used.

For multiple user transmission, coding of SU/MU [0], MU [1], MU [2] and MU[3] fields of VHT-SIG-A2 indicate the type of encoding for user u with USER_POSITION [u] = 0 , 1, 2 and 3, respectively.

The PHY layer decoder should use LDPC for decoding the C-PSDU for the respective user if the bit for its C-PSDU is 1. Decoding the BCC is used otherwise.

If VHT-SIG-B is invalid, the PHY shall issue the error condition PHY-RXEND.indication (FormatViolation) primitive. If the field VHT-SIG-B is decoded and the CRC field of VHT-SIG-B is activated and valid, a PHY-RXSTART.indication (RXVECTOR) primitive is issued.

In case of loss of signal at the reception before the end of the PSDU reception, the error condition PHY-RXEND.indication (CarrierLost) must be reported to the MAC. After waiting the end of the PSDU, PHY sends a PHY-CCA.indication (IDLE) primitive and return to the inactive state of reception. PSDU received bits are assembled into bytes and presented to the MAC using a series of PHY-DATA.indication (Data) primitives. After receiving the final bit of the last byte PSDU, the receiver must be return to the inactive state of reception. A PHY-RXEND.indication (NoError) primitive is issued upon entry to the inactive state of reception.

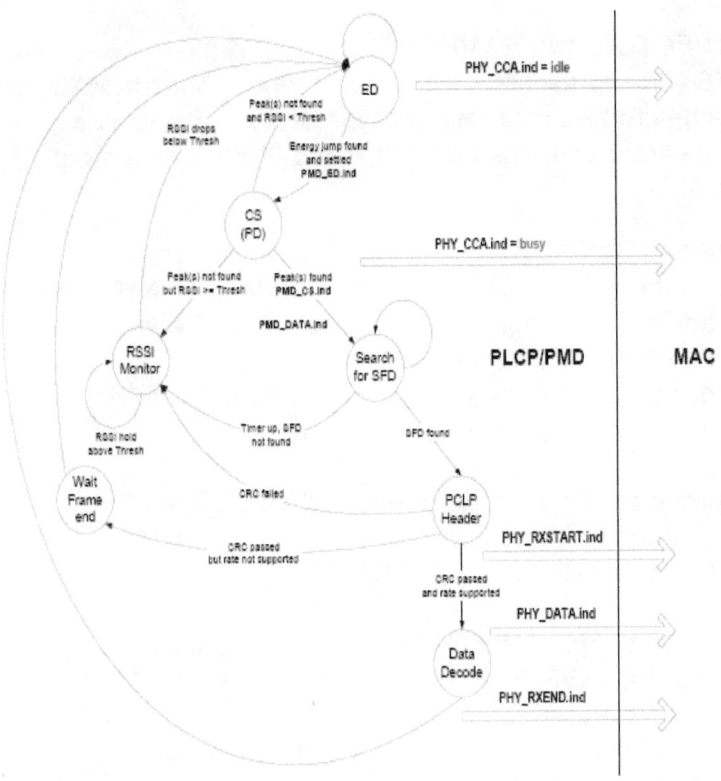

Figure 57 PHY/MAC interactions in reception

Whereas the PHY-SAP primitives are common for all 802.11 PHY layers, PMD-SAP primitives are specific to each interface.

4.3.5 *TXVECTOR* AND *RXVECTOR* PARAMETERS

The TXVECTOR represents a list of parameters that the MAC sub layer provides to the local PHY entity in order to transmit a PSDU. This vector contains both PLCP and PHY management parameters.

TXVECTOR and RXVECTOR parameters are set by the parameters of the TX-VECTOR sent in the PHY-TXSTART.request primitive and/or in the parameter RXVECTOR in the PHY primitive PHY-RXSTART.indication.
Actually there are many, many possible variations of PHY formats.

The PHY provides an interface to the MAC through the generic PHY service interface. The interface includes TXVECTOR, RXVECTOR, and PHYCONFIG_VECTOR.

The TXVECTOR supplies the PHY with transmit parameters per PPDU . Using the RXVECTOR, the PHY informs the MAC of the received PPDU parameters. Using the PHYCONFIG_VECTOR, the MAC configures the PHY for operation, independent of transmission or reception frame .

The FORMAT parameter of TXVECTOR determines the overall structure of the PPDU and includes:
- Non-HT format (NON_HT), based on OFDM.
- HT-mixed format (HT_MF).
- HT-Greenfield format (HT_GF).
- VHT (802.11ac) format.

Actually TXVECTOR and RXVECTOR are not used in Linux implementation of 802.11ac, but the PHY preamble is simply another transcription of those two pieces of information. This behavior mirrors the way the MLME semantics transpire in the MAC control header.

Parameter	Condition	Value
FORMAT		The FORMAT parameter of TXVECTOR determines the overall structure of the PPDU and includes: • Non-HT format (NON_HT), based on OFDM. • HT-mixed format (HT_MF). • HT-greenfield format (HT_GF). • VHT (802.11ac) format.
NON_HT_MODULATION	FORMAT is NON_HT	Indicates the format type of the transmitted non-HT PPDU.
	FORMAT is NON_HT	Indicates the estimated format type of the received non-HT PPDU.
L_LENGTH	FORMAT is NON_HT	Indicates the length of the PSDU in bytes
	FORMAT is HT_MF	Indicates the value in the Length field of the L-SIG
	FORMAT is HT_GF	Not used
	FORMAT is VHT	Not used
L_DATARATE	FORMAT is NON_HT	Indicates the data rate used to transmit the PSDU in Mb/s.
	FORMAT is HT_MF	Indicates the data rate value that is encoded in the L-SIG Rate field.
	FORMAT is HT_GF	Not used
	FORMAT is VHT	Not used
LSIGVALID	FORMAT is HT_MF	True if L-SIG Parity is valid. False if L-SIG Parity is not valid
	FORMAT is VHT	Not used
	else	Not used
SERVICE	FORMAT is NON_HT or HT_MF or	Scrambler initialization, 7 null bits + 9 reserved null bits

	HT_GF		
	FORMAT is VHT	Not used	
TXPWR_LEVEL		This is used to indicate which of the available TxPowerLevel attributes (values from 1 to 8) defined in the MIB shall be used for the current transmission.	
RSSI		This parameter is a measure by the PHY of the power observed at the antennas used to receive the current PPDU. RSSI shall be measured during the reception of the PLCP preamble.	
PREAMBLE_TYPE		Not used	
MCS		The modulation and coding scheme used in the transmission of the packet.	
REC_MCS		The receiver recommended MCS	
CH_BANDWIDTH	FORMAT is HT_MF or HT_GF	Indicates whether the packet is transmitted using 40 MHz or 20 MHz channel width. Enumerated type: * HT_CBW20 for 20 MHz and 40 MHz upper and 40 MHz lower modes * HT_CBW40 for 40 MHz	
	FORMAT is NON_HT	Enumerated type: * NON_HT_CBW40 for non-HT duplicate format * NON_HT_CBW20 for all other non-HT formats	
CH_OFFSET		Indicates which portion of the channel is used for transmission. Enumerated type: * CH_OFF_20 indicates the use of a 20	

		MHz channel (that is not part of a 40 MHz channel). * CH_OFF_40 indicates the entire 40 MHz channel. * CH_OFF_20U indicates the upper 20 MHz of the 40 MHz channel * CH_OFF_20L indicates the lower 20 MHz of the 40 MHz channel.	
LENGTH	FORMAT is VHT	Not used	
	FORMAT is HT_MF or HT_GF	Indicates the length of an HT PSDU in the range of 0 to 65 535 octets. A value of 0 indicates a NDP that contains no data symbols after the HT preamble	
	Other cases	Not used	
SMOOTHING	FORMAT is HT_MF or HT_GF	Indicates whether frequency-domain smoothing is recommended as part of channel estimation. Enumerated type: * SMOOTHING_REC indicates that smoothing is recommended. * SMOOTHING_NOT_REC indicates that smoothing is not recommended.	
	FORMAT is VHT	Not used	
AGGREGATION	FORMAT is HT_MF or HT_GF	Indicates whether the PSDU contains an A-MPDU. Enumerated type: * AGGREGATED indicates this packet has A-MPDU aggregation. * NOT_AGGREGATED indicates this packet does not have A-MPDU aggregation.	

	FORMAT is VHT	Not used	
STBC	FORMAT is HT_MF or HT_GF	Indicates the difference between the number of space-time streams (Nsts (# of SS)) and the number of spatial streams (Nss) indicated by the MCS as follows: * 0 indicates no STBC (Nsts (# of SS) = Nss). FORMAT is HT_MF or HT_GF * 1 indicates Nsts (# of SS) – Nss = 1 . * 2 indicates Nsts (# of SS) – Nss = 2 . * Value of 3 is reserved.	
	FORMAT is VHT	Indicates whether or not STBC is used. * 0 indicates no STBC (Nsts (# of SS)= Nss in the Data field). * 1 indicates STBC is used (Nsts (# of SS)=2 * Nss in the Data field).	
	Other	Not used	
FEC_CODING	FORMAT is HT_MF or HT_GF or VHT	Indicates which FEC encoding is used. Enumerated type: * BCC_CODING indicates binary convolutional code. * LDPC_CODING indicates low-density parity	
	Other	Not used	
GI_TYPE	FORMAT is HT_MF or HT_GF or VHT	Indicates whether a short guard interval is used in the transmission of the packet. Enumerated type: * LONG_GI indicates short GI is not used	

		in the packet. * SHORT_GI indicates short GI is used in the packet.	
	Other	Not used	
NUM_EXTEN_SS	FORMAT is HT_MF or HT_GF	Indicates the number of extension spatial streams that are sounded during the extension part of the HT training in the range of 0 to 3.	
	Other cases (including VHT)	Not used	
ANTENNA_SET	FORMAT is HT_MF or HT_GF	Indicates which antennas of the available antennas are used in the transmission. The length of the field is 8 bits. A 1 in bit position n, relative to the LSB, indicates that antenna n is used. At most 4 bits out of 8 may be set to 1. This field is present only if ASEL is applied.	
	Other cases (including VHT)	Not used	
N_TX	FORMAT is HT_MF or HT_GF or VHT	The N_TX parameter indicates the number of transmit chains.	
	Other cases	Not used	
EXPANSION_MAT	EXPANSION_MAT_TYPE is COMPRESSED_SV	Contains a set of compressed beam forming feedback matrices. The number of elements depends on the number of spatial streams and the number of transmit chains.	
	EXPANSION_MAT_TYPE is NON_COMPRESSED_SV	Contains a set of noncompressed beam forming feedback matrices where Nst is the total number of sub-carriers, Nc is the number of columns, and Nr is the number of rows in each matrix.	

	EXPANSION_MAT_TYPE is CSI_MATRICES	Contains a set of CSI matrices. The number of complex elements is N ST × N r × N c where N ST is the total number of sub-carriers, N c is the number of columns, and N r is the number of rows in each matrix.	
EXPANSION_MAT_TYPE	EXPANSION_MAT is present	Enumerated type: COMPRESSED_SV indicates that EXPANSION_MAT is a set of compressed beam forming feedback matrices. NON_COMPRESSED_SV indicates that EXPANSION_MAT is a set of noncompressed beam forming feedback matrices. CSI_MATRICES indicates that EXPANSION_MAT is a set of channel state matrices.	
	Otherwise	Not used	
CHAN_MAT	CHAN_MAT_TYPE is COMPRESSED_SV	Contains a set of compressed beam forming feedback matrices based on the channel measured during the training symbols of the received PPDU. The number of elements depends on the number of spatial streams and the number of transmit chains.	
	CHAN_MAT_TYPE is NON_COMPRESSED_SV	Contains a set of noncompressed beam forming feedback matrices based on the channel measured during the training symbols of the received PPDU. The number of complex elements is Nst × Nr × Nc where Nst is the total number of sub-carriers, Nc is the number of columns, and N r is the number of rows in each matrix.	
	CHAN_MAT_TYPE is CSI_MATRICES	Contains a set of CSI matrices based on the channel measured during the training symbols of the received PPDU. The number of complex elements is Nst × Nr × Nc where Nst is the total number of sub-carriers, Nc is the number of columns, and Nr is the	

		number of rows in each matrix.
	Other	Not used
RCPI		Is a measure of the received RF power averaged over all the receive chains in the data portion of a received frame.
SNR	CHAN_MAT_TYPE is CSI_MATRICES	Is a measure of the received SNR per chain. SNR indications of 8 bits are supported. SNR shall be the decibel representation of linearly averaged values over the tones represented in each receive chain .
	CHAN_MAT_TYPE is COMPRESSED_SV or NON_COMPRESSED_SV	Is a measure of the received SNR per stream. SNR indications of 8 bits are supported. SNR shall be the sum of the decibel values of SNR per tone divided by the number of tones represented in each stream.
NO_SIG_EXTN	FORMAT is NON_HT and NON_HT_MODULATION is ERP-OFDM, DSSS-OFDM, or NON_HT_DUPOFDM	Indicates whether signal extension needs to be applied at the end of transmission. Boolean values: * true indicates no signal extension is present. * false indicates signal extension may be present depending on other TXVECTOR parameter.
	Other	Not used
TIME_OF_DEPARTURE_REQUESTED		Enumerated type: True indicates that the MAC entity requests that the PHY PLCP entity measures and reports time of departure parameters corresponding to the time when the first frame energy is sent by the transmitting port. False indicates that the MAC entity requests that the PHY PLCP entity neither measures nor reports time of departure parameters.
RX_START_OF_FRAME_OFFSET		0 to $2^{32} - 1$. An estimate of the offset (in 10 ns units) from the point in time at which the

		start of the preamble corresponding to the incoming frame arrived at the receive antenna port to the point in time at which this primitive is issued to the MAC.

Figure 58 TXVECTOR and RXVECTOR parameters

4.3.6 802.11AC PHY PREAMBLE.

The PHY frame consists roughly of three parts.

- The first (Training Field) of the frame allows the receiver to synchronize with the transmitter. It is more or less complex according to the needs that range from compatibility with older PHY layers, to the formation of MIMO beams. This first part is named the preamble it is sent at 6 Mbits/sec.

- The second part (SIG) of the frame vehicles additional configuration information destined for the receiver and the third consists of the data to be transmitted. The first part has a field format and modulation method which is in common to all 802.11 amendments. The PLCP header is encoded at a fixed data rate, so that all receiving stations either of last generation or of legacy technologies, can decode the information.

- The third part of the frame is transmitted at the maximum rate that allows the channel. It is filled in with MAC's A-MPDU.

The OFDM PLCP preamble is transmitted at six Mbps. The PPDU frame payload is encoded at a variable rate of data that is specified in the SIG field.

802.11ac PHY layer design closely follows the PHY layer of the 802.11n amendment.

The preamble of the PHY frame consists of the following fields in this order:
- The classical learning short preamble (STF): Detection of frame beginning, determination of the MCS, of the initial frequency and time synchronization. Basically it means detecting that a carrier exists and synchronizing the local oscillator with it.
- The classic long duration training preamble (LTF) is used for fine synchronization of the frequency channel

estimation and time. It aims at detecting the OFDM clock
.

- Classic Signal Field(L-SIG): It indicates the length of the payload.
- VHT-SIG-A: replace the HT-SIG field of 802.11n, it contains 802.11ac PHY single-user and some MU settings.
- VHT-STF: similar to the standard 802.11n HT-STF, allows the adjustment of the automatic gain control.
- VHT-LTF: similar to HT-LTF 802.11n, which is used for the estimation of channel.
- VHT-SIG-B: a new field in 802.11ac, containing additional parameters.

The fields L-STFs, L-SIG, VHT-SIG-A, VHT-STF are 20 MHz waveforms replicated on each sub-channel of the 80 MHz channel for making them look like legitimate 802.11a channels.

L-STF	L-LTF	L-SIG	VHT-SIG-A	VHT-STF	VHT-LTF	...	VHT-LTF	VHT-SIG-B	Data

Figure 59 PPDU 802.11ac format

The VHT-SIG-A, VHT-STF, VHT-LTF, and VHT-SIG-B fields exist only in 802.11ac PPDU. In a NDP frame (Null Data frame for sounding purposes), the Data field is not present. The number of symbols in the VHT-LTF field, NVHTLTF is determined by the total number of space-time streams across all users being transmitted in the VHT PPDU.

The joint 802.11ac preamble format has the following characteristics:

- Support for 802.11a. The 802.11ac MF preamble is designed so that an 802.11a device can inter-operate with an 802.11ac device.

214

- 802.11n support. The 802.11ac MF preamble is designed so that an 802.11n device can inter-operate in an 802.11ac device within the 5 GHz band.

The number of sub-carriers is the same as those of 802.11n LTF-L and L-SIG fields at 20 MHz in each 20 MHz sub channel. VHT-LTF fields and the field positions of data symbols in the 20 and 40 MHz channels are the same as those for 802.11n HT-LTF and HT-DATA for 20 to 40 MHz channels.

4.3.6.1 PHY training Fields

The training field purpose is to make it possible for the receiver to get information about the carrier frequency and phase, and about the OFDM symbols clock.

The radio channel is changing over time. A channel estimation process is therefore necessary for each incoming frame so as to cope with these changes. It is necessary not only to estimate the parameters of the channel, but also to estimate the parameters of each of the OFDM sub carriers. A particular sequence of symbols, which is known by all the stations, is used for this purpose. Channel estimation makes it possible to obtain a matrix that is used to level the rest of the frame. Estimation of frequency offset and clock synchronization can be refined through the channel estimate.

There are three mandatory formats in the standard for PLCP PPDU, but the first two are obsolete because specific to (802.11b) DSSS modulation.

- Long preamble for DSSS
- Short preamble for DSSS
- Preamble for OFDM

802.11g and 802.11a introduced the use of OFDM respectively in 2.4 GHz and 5 GHz bands. OFDM carries symbols on many different sub carriers. The synchronization is crucial and must precede any other treatment. Indeed, a misinterpretation of the received signal can lead to a wrong decoding of data symbols. In IEEE 802.11a and 802.11g standards, it is partially done using the short duration training field (STF) field that is specific to OFDM but it is not sufficient to enable symbol decoding. STF makes it possible for the receiver to learn the phase and frequency of the carrier but it can learn nothing about the OFDM symbols clock. For that kind of learning, it needs a sequence of known in advance OFDM symbols but decoding OFDM needs knowledge of carrier properties, STF is therefore the first field in PLCP preamble. The short preamble also performs: Detection of beginning of frame, determination of the MCS for the LTF field.

The STF is followed by the LTF field, which duration is of 8 micro seconds. LTF is an acronym for "Long training Field" and consists of two long sequences of OFDM symbols. As no radio channel is perfect, LTF is also used for the channel estimation, i.e. to gain knowledge about the various distortions and heterogeneity in the channel. The two long sequences of OFDM symbols last 3.2 ms each, and are protected by a guard interval (GI) of 1.6 micro seconds. These sequences are designed to make the process more robust to noise and estimation errors.

To recap the "classic" OFDM preamble is sent at 6 Mbits/sec and contains the following fields:

- STF: 10 sequences of short learning which is used for the convergence and MCS learning, the acquisition of synchronization, and the coarse acquisition of frequency for a duration of 0.8 micro seconds each (10 * 0.8 = 8 micro seconds).
- LTF: 2 long learning sequences used for channel estimation and acquiring fine frequency of a length of 3.2 micro seconds per OFDM symbol.
- A guard (IM) of 1.6 micro seconds interval after each symbol.

As 802.11n and 802.11ac aim at being compatible with 802.11a, the new PHY frame is enclosed in a legacy PHY frame.

The STF and LTF fields stay but new synchronization fields are added.
In addition from a 802.11a point of view, a given 20MHz channel used by a 802.11ac station looks like a 802.11a channel up to the sub-carrier level, including the pilot sub-carriers even in the data part of the 802.11a PHY frame. Sometimes it goes as far as spoofing a 6Mbits/sec MCS, for example for the L-SIG field.

The 802.11n added HT-STF and HT-LTF fields and 802.11ac adds VHT-STF and VHT-LTF fields.
The 802.11ac amendment introduces a specific LTF field for very high-speed (VHT–LTF). The principle is similar to that used for HT-LTF, except that each VHT – LTF on a different spatial stream has a different pre-coding to introduce diversity between spatial streams.

A new STF field at very high speed (VHT-STF) is also defined, it is derived from the HT-LTF 802.11n field.

VHT-STF definition
The field frequency sequence used to construct the 802.11ac VHT-STF in a 20 MHz transmission is identical to the L-STF.
In a transmission at 40 MHz, the 802.11ac VHT-STF is built from the 20 MHz version by duplication, offset frequency and rotation of 90° of the higher 20 MHz sub carriers.
In a transmission at 80 MHz, the 802.11ac VHT-STF is built from the 20 MHz version by reproducing each band of 20 MHz, shifting the frequency, and by applying appropriate phase rotations for each 20 MHz sub-band.
The same procedure is used for a transmission at 160 MHz.
For non-contiguous transmissions using two 80 MHz segments, each segment of 80 MHz must use the 80 MHz VHT-STF model.

VHT-LTF definition
802.11ac is more sensible to errors in signal distortions than previous generation due to higher MCS and increased use of MIMO. This long training fields consist of one, two, four, six or

eight OFDM symbols, that are required for demodulation of the 802.11ac data field, depending on the number of spatial streams. Contrary to previous fields, VHT-LTF field do not spread each symbol on each 20MHz channel, it uses the full width of the 802.11ac channel. Nevertheless VHT-LTF symbols have the same number of pilot and data sub carriers per 20MHz channel as in a 20MHz 802.11a channel. Pilot sub-carriers are important to track the phase of the OFDM modulation.

The matrix P of look-up table of VHT-LTF is applied to all sub-carriers in VHT-LTF symbols, except for pilot sub carriers. Instead, a repeat of row matrix R of look-up table is applied to all pilot sub carriers in VHT-LTF symbols .
The matrix R row repetition operation has the same dimensions as the matrix P (Nsts x NLTF), with all the rows of the matrix R identical to the first row of the matrix P of the corresponding dimension.
This simple solution helps in tracking phase and frequency shift.

4.3.6.2 Signalization (SIG) field
The legacy signal Field (L-SIG) is placed just after the preamble. It contains two OFDM symbols. A parity bit checks its integrity. From the point of view of a 802.11a station examining a 20MHz channel which is a part of a 802.11ac PHY frame, this field looks like a data field, with the correct number of sub-carriers of data and pilot type.

The signal field contains:
- Information on the information throughput
- On the length (12-bit)
- Other fields including the identity of the receiving station field.

This header follows the PLCP preamble and provides information about MCS, channels, and addresses. Indeed, the receiver must know precisely what type of modulation and coding are applied by the issuer on data symbols. The amount of transmitted data is also important. If special options are enabled and required for a correct decoding, they shall be noted in the PLCP header. If it is misinterpreted, the data contained in the frame will not be recovered. This field is thus often protected by an integrity check. Signalization in the band is contained in the PLCP header.

This signal field allows the transmitter to send information to the receiver, because it needs information to correctly decode the frame as there are many options and parameter values.

This field, that exist since the beginnings of 802.11 had a first important evolution with 802.11n (which introduced the HT-SG fields) then another with 802.11ac (which introduced the VHT-SIG-A and VHT-SIG-B).

It is this VHT-SIG field, which allows a station that is listening on the channel, to know if the rest of frame is sent to it. With later amendments this field reflects a more complex identity, for example it could show to which MU-MIMO group the station belongs.

The 802.11ac SIG field is divided into two fields. The VHT-SIG-A field replaces the HT-SIG field of the 802.11n standard and contains settings common to all stations concerned by this frame, for example in case of group addressing. The field VHT-SIG-B contains settings specific to each station. It should be noted that when MU-MIMO is used, all the VHT-SIG-A fields are differently pre-coded to ensure the diversity of the paths, as there is one specific VHT-SIG-A field per path. The beam forming begins only after VHT-SIG-B when the station has everything it needs. The preamble is readable by all stations up to VHT SIG B included.
The highest specified structure (of L-STF up to VHT-SIG-B) is repeated for each channel.

The legacy Signal field (L-SIG) of older devices, indicates the length of the payload, the VHT-SIG-A symbol replaces the HT-SIG field of 802.11n, containing 802.11ac PHY for single-user and some multi-user settings. VHT-STF in a similar way to the

802.11n HT-STF, allows the adjustment of the automatic gain control. The fields of the PHY frame up to and including the VHT-STF field consist of 20 MHz channels. This is replicated in the slot adjacent channels.

VHT-SIG-A definition

VHT-SIG-A1

B0-B1	Channel Width Set to: - 0 for 20 MHz, - 1 for 40 MHz, - 2 for 80 MHz, - 3 for 160 MHz and 80+80 MHz
B2	Reserved, set to 1.
B3	STBC - For a VHT SU PPDU: Set to 1 if space time block coding is used and set to 0 otherwise. - For a VHT MU PPDU: Set to 0.
B4-B9 Group ID	Set to the value of the TXVECTOR parameter GROUP_ID. - A value of 0 or 63 indicates a VHT SU PPDU. - otherwise it indicates a VHT MU PPDU.
B10-B21 Nsts/Partial AID	For a VHT MU PPDU: Nsts is divided into 4 user positions of 3 bits each. Each user position contains the number minus one of space time streams
	For a VHT SU PPDU, - Bits 10 to 12 indicates the number minus one of space time stream - Bits 13 to 21, Partial AID: Set to the value of the TXVECTOR parameter PARTIAL_AID. Partial AID provides an abbreviated indication of the intended recipient(s) of the PSDU
B22	TXOP_PS_NOT_ALLOWED: Set to 0 by an access point if it allows non-AP VHT STAs in TXOP power save mode to enter Doze state during a TXOP. Set to 1 otherwise.

Figure 60 VHT-SIG-A1 definition

VHT-SIG-A2

B0 Short GI		
	Set to 0 if short guard interval is not used in the Data field.	
	Set to 1 if short guard interval is used in the Data field.	
B1 Short GI NSYM Disambiguation		
	Set to 1 if short guard interval is used and NSYM mod 10 = 9, otherwise set to 0.	
B2 SU/MU[0] Coding		
	For a VHT SU PPDU, B2 is set to 0 for BCC, 1 for LDPC	
	For a VHT MU PPDU,	if the MU[0] Nsts field is nonzero, then B2 indicates the coding used for user u with USER_POSITION[u] = 0; set to 0 for BCC and 1 for LDPC.
		If the MU[0] Nsts field is 0, then this field is reserved and set to 1.
B3 LDPC Extra OFDM Symbol	Set to 1 if the LDPC PPDU encoding process (if an SU-PPDU), or at least one LDPC user's PPDU encoding process (if a VHT MU PPDU), results in an extra OFDM symbol (or symbols)	
	Set to 0 otherwise.	
B4-B7 SU VHT-MCS/MU[13] Coding	For a VHT SU PPDU:	VHT-MCS index
	For a VHT MU PPDU:	
		If the MU[1] Nsts field is nonzero, then B4 indicates coding for user u with USER_POSITION[u] = 1: set to 0 for BCC, 1 for LDPC.
		If the MU[1] Nsts field is 0, then B4 is reserved and set to 1.
		If the MU[2] Nsts field is nonzero, then B5 indicates coding for user u with

		USER_POSITION[u] = 2: set to 0 for BCC, 1 for LDPC.
		If the MU[2] Nsts field is 0, then B5 is reserved and set to 1.
		If the MU[3] Nsts field is nonzero, then B6 indicates coding for user u with USER_POSITION[u] = 3: set to 0 for BCC, 1 for LDPC.
		If the MU[3] Nsts field is 0, then B6 is reserved and set to 1.
		B7 is reserved and set to 1
B8 Beam-formed	For a VHT SU PPDU:	Set to 1 if a beam-forming steering matrix is applied to the waveform in an SU transmission
		set to 0 otherwise.
	For a VHT MU PPDU:	Reserved and set to 1
B9	Reserved and set to 1	
B10-B17	CRC	
B18-B23	Tail Used to terminate the trellis of the convolutional decoder. Set to 0.	

Figure 61 VHT-SIG-A2 definition

VHT-SIG-B definition

VHT-SIG-B consists of 26 bits for a PPDU with 20 MHz, 27 bits for transmission at 40 MHz and 29 bits for transmission to 80 MHz.

- For the PPDU 40 MHz and 80 MHz, 802.11ac VHT-SIG-B bits are repeated.

- In a PPDU to 160 MHz, the 80 MHz to 802.11ac VHT-SIG-B is repeated twice in frequency.

VHT-SIG-B includes the fields listed below

Field	MU bit allocation			mono user bit allocation			Description
	20 MHz	40 MHz	80 MHz	20 MHz	40 MHz	80 MHz	
Length	16	17	19	17	19	21	length of useful data in PSDU in units of 4 bytes
MCS	4	4	4	-	-	-	
Reserved	0	0	0	3	2	2	All bits are at "one"
Tail	6	6	6	6	6	6	All bits are at "zero"

Total # bits	26	27	29	26	27	29	

Figure 62 802.11ac VHT-SIG-B fields

The size of the 'Length' field varies with the width of channel and mono-user versus multi-user to ensure that a consistent timing of 5.4ms is maintained.

SERVICE field

The SERVICE field is scrambled. It is shown below:

Bits	Field	Description
B0-B6	Scrambler Initialization	
B7	Reserved	
B8-B15	CRC	CRC calculated over a 802.11ac VHT-SIG-B (excluding tail)

Figure 63 SERVICE field

VHT-SIGB   Service Field

20 bits in 20MHz *21 (40MHz) / 23(80MHz) bits	Tail (6bit)	Scrambler Seed (7bit)	Rsvd (1 bit)	CRC (8 bit)

Figure 64 VHT-SIG-B and SERVICE field relationship

4.3.6.3 Data field

The "Data" field contains:

1. A service information
2. The PSDU
3. Other less important information

After the PHY preamble, there is the PHY data field. The first 16 bits are the service field. In 802.11ac, this field has changed to include a CRC to VHT-SIG-B. The change is due to the addition of MU-MIMO and the increase in the maximum number of bytes per packet, which in turn means that the length of the packet should be reported differently. The data is encoded and encrypted. It is followed by the constellation and space lookup table.

4.3.6.4 SIG fields in single user case

For a single user transmission, using the BCC, the coding process goes as follows. The MAC layer calculates Nsym and NPAD. The MAC layer is then filled to the last octet and indicates (using the TXVECTOR) the number of pad bits of the PHY to be added. After receiving the PSDU, the PHY adds bits 0-7, and then adds the NES bits at the back of each encoder.

For multi-user transmission using BCC encoding, the filling process goes as follows. The MAC layer calculates Nsym for each user separately. For each user, on the basis of the maximum number of symbols on all users, MAC fills them to the last byte limit. The number of bits of stuffing PHY added for each user is calculated as single user. For each user, encoding and flow analysis is done as single user.

4.3.6.5 SIG fields in multi-user case

The training headers (VHT-LTF) provide a means for the receiver to estimate the MIMO channel. In this way, each automatic control of gain and phase of channel of each sub-carrier used for the transmission of data can be properly estimated.

In addition, the first field in PLCP, which is duplicated by 20 MHz channel, is also sent without precoding (i.e. in an omnidirectional manner) in a manner to be understood by all stations.

While in reception phase, 802.11ac stations can determine if they are involved in the incoming frame, thanks to the signal field VHT-SIG-A. For this purpose, the group identifier (GID) and a partial association ID (AID partial) are inserted in the VHT-SIG-A field.

When stations associate to an access point (AP), they are assigned a 14-bit identity to help with monitoring and management functions. The partial AID on 9 bits is obtained by hashing these identities with the 48-bit of the base (BSSID) Service Set ID field to uniquely identify a station. However, when MU-MIMO is used the MU-MIMO stations need to know if they are part of the group currently served by the emitter station. So when MU-MIMO is used, up to 62 different possible groups are defined by the access point and transmitted to all the MU-MIMO Stations of the BSS. Thanks to the 6-bit group ID, stations can also know their order within the group.

4.3.7 MANAGEMENT OF MULTI-CHANNELS

The earlier 802.11 stations use channels that were about 20 MHz wide. 802.11a had even half and quarter channels of 10 and 5MHz.

Products based on 802.11n can use channels for 20 or 40 MHz wide or in the ISM band (2, 4 GHz) or UNII (5 GHz) which allows them to increase available throughput. In 802.11ac the stations must support 80 MHz wide channels and may be optionally as wide as 160 MHz or 80 MHz + 80 MHz.

Mechanisms for access to the media with aggregated channels have already defined in Amendment IEEE 802.11n for 40 MHz transmissions. In IEEE 802.11ac the principle is extended to 80 MHz to 160 MHz transmissions.

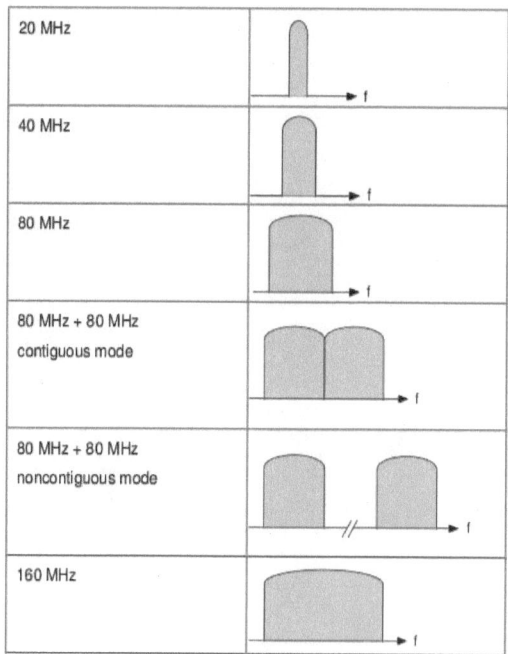

Figure 65 802.11ac aggregated channel

4.3.7.1 Management of channels in 802.11ac

However, interoperability means that a simple 802.11 channel at 20 MHz used by a legacy station can prevent a secondary channel to be used optimally by a 802.11ac station, even without addressing the subject of interference between channels. In 802.11, CCA is used to detect the presence of signals on the radio media and thus allows it to defer appropriately any transmission by the station which is implementing CCA, until the media becomes free.

The basic condition for a station using OFDM is to detect a valid signal at a level of-82 dBm. It should also detect any other signal at a level of-62 dBm. It is called the energy detection (ED). When 802.11n added the 40 MHz channel with a main channel at 20 MHz and a secondary channel at 20 MHz, the energy detection (ED) became necessary on the secondary channel due to the increased complexity of detection of a valid signal on the secondary channel.

However for a station to detect an interference on any secondary channel by another station, it implies that the signal is at a level higher than -62 dBm. This means that any other station that uses the same secondary channel as a primary channel in another BSS, is disadvantaged by 20 dB because it stops to use this primary channel as soon as the CCA detects an energy level higher than -82dBm, where the other station will stop only if it detects an energy higher than -62 dBm.

In 802.11ac, detection of a valid signal in the secondary channel is at a level of-72 dBm or - 69 dBm one bandwidth to improve the performance of the CEC channel side. In addition, it requires that a device for detecting a valid packet in secondary channels is done not only on the basis of the preamble to a package, but also in the center of the package.

If we assume that the 802.11ac has the primary channel a level of sensitivity of CCA of-82 dBm and the level of sensitivity of the secondary CCA or -62 dBm channel, then, if the power of the received signal at the level of the secondary station 802.11ac channel is greater than-82 dBm, but less than-62 dBm, station

802.11ac consider the secondary channel as inactive and will generate a signal to 80 MHz wide which can collide with another pass on the secondary channel.

The rules of transmission in the 802.11ac standard are as follows:

- The beginning of a packet transmission time is solely based on the CCA's main channel result.
- Controls of 802.11n stations transmit only if the main channel has been idle for a time equal to a DIFS plus the CW interval.

If the main channel has been idle for this period of time and if the secondary channel has also been inactive for an inter frame spacing, the station may transmit a signal immediately after the expiration of the time counter.

However, if the secondary channel was active during this interval, the station has two choices:

- Static operation at full channel width: The station can attempt to gain access to the 40 MHz channel by restarting the attempt to access the channel with a new value chosen at random for the counter "backoff".
- Dynamic operation at the best channel width: The station can transmit a signal only on the 20 MHz primary channel.

The differences between the two types of operations are as follows, but will be detailed elsewhere:

- (1) Sensitivity of CCA (Clear Channel assessment): the level of sensitivity of CCA on the primary channel is much lower than that of the secondary channels. The CCA's primary channel sensitivity level is -82 dBm for a signal at 20 MHz while the level of sensitivity of CCA on the secondary channels is - 62 dBm. The level of sensitivity of-62 dBm of the CCA can be accomplished by a simple energy detection system and so it is not necessary to implement CCA for the secondary channels.
- (2) Frame NAV (Network Allocation Vector): 802.11 frames received in the primary channel are decoded and

the duration of MAC header field is used to set the NAV value to a nearby station to not cause collision with the current transmissions.

4.3.7.2 Access to aggregated channels in 802.11ac

If an 802.11ac station wins a TXOP through the Distributed Enhanced mechanism (EDCA) on the main channel, it therefore can send a PPDU at 80 MHz channel width if all other channels have been inactive for at least a point coordination function inter-frame (PIFS). If at least one of the secondary channels has been active during a PIFS, the station should start its "backoff" count again, or use the TXOP obtained for 20 or 40 MHz PPDU. In the IEEE 802.11n standard, two 20 MHz channels can be linked to form a 40 MHz bandwidth channel. As an evolution of the 802.11n standard, IEEE 802.11ac adds 80 MHz, 160 MHz contiguous and non-contiguous channels. This last mode "non-contiguous" is also called 80 + 80 MHz channel and is to use two non-adjacent channels of 80 MHz (2 * 80 MHz).

The idea of adjacent non-channels binding had yet been proposed unsuccessfully during the development of 802.11n. Indeed, in 802.11n, 40 MHz channels consist of two channels of 20 MHz adjacent (a primary and the other secondary). These channels are not bound but partially overlap. 80 MHz channels are thus formed by two contiguous to 40 MHz channels, in which one of the 20 MHz channel is the primary sub-channel and others are secondary sub-channels (Actually, secondary, tertiary and quaternary channels). In North America, the several channels have been added in order to have 8 channels at 80 MHz, instead of five. The FTC is discussing opening more spectrum, in UK OFCOM made a public request for comment on this subject. Otherwise this would be discussed in 2015 at a world telecommunication conference.

If the channels at 160 MHz are used they are formed with two channels of 80 MHz each that can be either contiguous or non-contiguous. In the 5 GHz band there is sufficient bandwidth only for two contiguous channels to 160 MHz. This is the reason why non-contiguous channels at 160 MHz have been implemented, allowing numerous combinations of channels to 80 MHz.

4.3.7.3 Coexistence of channels a 802.11a/n and of 802.11ac channels

The coexistence of 802.11a/n networks and 802.11ac networks, the fact that the width of the channel widens and the 802.11 networks radio environment becomes denser because of the always increasing number of wireless devices, makes it more difficult for An 802.11ac access point to find a 80 MHz channel free.

As a result, a network 802.11ac is very likely to share a radio 80 MHz channel with 802.11 a/n legacy technology stations, co-located and operating on 20 or 40 MHz channels. The throughput of the network 802.11ac can strongly degrade if a co-located 802.11n network operates on a secondary channel of the 802.11ac network. This may be due, for example, to the fact that with frame aggregation, just one problem on a secondary channel (which is preempted and not subject to conditional access by CSMA/CA) invalidates all of the aggregated frame must be then send again. This is an example where a problem on the PHY layer impacts the MAC layer. Thus, it is important to use the frequency resource effectively by changing the sizes of channels between 20, 40 and 80 MHz.

4.3.7.4 Selection of primary channel

The position of the primary channel in 20 MHz channels can have an impact on the 802.11ac throughput when stations use the dynamic management of channel width.

An 802.11ac access point can put the primary channel 20 MHz at the beginning, at the end or in the middle of the 80 MHz channel.

First channel	Second channel	Third channel	Fourth channel
P	S1	S2	S3
S1	P	S2	S3
S3	P	S1	S2
S3	S2	P	S1
S3	S2	S1	P

When the primary channel is placed at the beginning or the end of the 80 MHz channel, the PPDU at 40 MHz may be transmitted only on the P-primary channel and on the adjacent S1 secondary, i.e. (P + S1):

First channel	Second channel	Third channel	Fourth channel
P	S1	S2	S3
S3	S2	S1	P

If S1 and one of the other two channels are busy, the station must withdraw at 20 MHz and only use the first 20 MHz for the transmission of the PPDU:

First channel	Second channel	Third channel	Fourth channel
P	XX	S2	XX

To

First channel	Second channel	Third channel	Fourth channel
P	XX	--	XX

However, if the primary channel is set in the middle of the 80 MHz channel, the station 802.11ac has one more option to pass PPDU to 40 MHz:

First channel	Second channel	Third channel	Fourth channel
S1	P	S2	S3
S3	P	S1	S2
S3	S2	P	S1

For example, when the S2 secondary channel is busy, it can use (S1 + P) to 40 MHz transmissions:

First channel	Second channel	Third channel	Fourth channel
S1	P	S2	S3
S3	P	S1	S2
S3	S2	P	S1

To

First channel	Second channel	Third channel	Fourth channel
S1	P	XX	--
--	P	S1	XX
--	--	P	S1

When S1 is busy, it can always use (P + S2) to 40 MHz, which will improve the flow of the 802.11ac station when the secondary channels are busy:

First channel	Second channel	Third channel	Fourth channel
S1	P	S2	S3
S3	P	S1	S2
S3	S2	P	S1

To

First channel	Second channel	Third channel	Fourth channel
XX	P	S2	--
S3	P	XX	--
--	S2	P	--

However this created a possibility of overlap (P + S2), across the border from 40 MHz channel with another 802.11n station 40 MHz operating in (S2 + S3):

From the point of view of the 802.11n station

First channel	Second channel	Third channel	Fourth channel
--	--	P	S1

From the point of view of the 802.11ac station

First channel	Second channel	Third channel	Fourth channel
XX	P	S2	--

If S2 is also the secondary station 802.11n 40 MHz channel, secondary channels of the 802.11ac and those of the 802.11n station overlap with each other and the 802.11n station may not be able to decode the packets received on its secondary channel S2, because 40 MHz (P + S2) transmissions can collide with the 802.11n station.

From the point of view of the 802.11n station

First channel	Second channel	Third channel	Fourth channel
--	--	S1	P

From the point of view of the 802.11ac station

First channel	Second channel	Third channel	Fourth channel
XX	P	S2	--

This problem can be solved by allowing the 802.11ac to 40 MHz in P and S2 transmissions only if S2 corresponds to the main channel of the station from 40 to 80 MHz2.802.11n.

4.3.7.5 reinforced Protection

With the possibility of linking up to eight 20 MHz channels in
IEEE 802.11ac, it becomes much harder to avoid overlap
between neighboring BSS, despite the large number of 20 MHz
channels available. To resolve this problem, the mechanism of
coexistence between channels is improved through three
techniques.

- First, evaluation of free secondary channels is
  strengthened in order to improve the detection of the
  signal.
- On the other hand, the basic RTS and CTS mechanism to
  gain authorization before emitting is modified with a
  mechanism to improve the functioning of the channel
  dynamic width. The station sending RTS frames can send
  data only on the aggregated channels acknowledged by
  CTS frames (with RTS and CTS frames duplicated on
  each 20 MHz channel).
- Third, notification frames were put in place to coordinate
  the bandwidth in case of too frequent noise.

4.4 MISCELLANEOUS PLCP FUNCTIONS

4.4.1 THROUGHPUT ADAPTATION

Any transmission channel is imperfect and there is a relationship between the noise on this channel and its maximum information throughput (Shannon law). It is therefore interesting to try to maintain the information throughput just below the upper limit for maximum information throughput. This implies to develop an indicator to either show that information throughput must be lowered because there are too many errors in transmission; or on the contrary show that increase the information throughput is needed because a significant margin exists. It is important to tell the difference between noise and collision. The specification allows a station to transmit even if it detects a significant level of noise (-62dBm); on the other hand, it is prohibited to transmit if it detects an existing valid 802.11 carrier on the media, even if it is at a relatively low level (-82dBm). However if we consider a single frame, it is the throughput limited by the signal/noise ratio (SNR) which is important, collision detection does not matter because by definition the frame may not be issued unless the channel was free. An issue of the rate adaptation will be to estimate the noise level to derive the maximum information throughput rate.

4.4.2 CLEAR CHANNEL ASSESSMENT

The detection of carrier in 802.11 consists of two functions separate and distinct, Clear Channel Assessment (CCA) and the Network Allocation Vector (NAV). From a high-level perspective, the CCA is the PHY layer measure energy received on the interface. NAV is the virtual carrier mechanism that is used by stations to reserve the media for the higher priority frames that must absolutely be transmitted just after the current frame, for example aggregated frames. NAV is a MAC-level device, so it is studied in the MAC section.

When the MAC layer receives a packet to be transmitted, it requests the PHY layer to check is the channel is clear i.e. to do a CCA. If the CCA indicates a free channel, the MAC layer asked the PHY layer begin to transmit the packet. However if CCA indicates that a channel is occupied, the MAC layer waits for a certain period (called "backoff") and restarts the process.

Clear Channel Assessment consists of two functions, energy detection and the detection of carrier (Carrier Sense CS).

- CS with a threshold of-82 dBm
- ED with a threshold of-62 dBm

4.4.2.1 CCA and aggregated channels

802.11ac stations have an improved detection of electromagnetic activity in secondary channels so that they can thus better avoid collisions.

CCA should allow the detection and the report on all possible combinations of use of channels and signaling in an inactive radio bandwidth.

An 802.11ac device must provide a CCA per channel, for all the channels that, the device is capable of operating. Detecting a valid signal in secondary channel is harder than in primary channel

- Because the STA always transmits in the primary channel, it only needs to detect start of packet in primary channel

- Because a secondary transmission may begin while a primary channel transmission is in progress, a STA must be able to detect signal in middle of a packet on secondary channel

Primary channel

* valid 802.11 signal: -82 dBm
* Energy detect: -62 dBm

Secondary channels

* valid 802.11 signal:
  - 20 MHz: -72 dBm
  - 40 MHz: -72 dBm
  - 80 MHz: -69 dBm
* Energy detect: -62 dBm

4.4.2.2 Energy detection (ED)

The detection of energy (ED) refers to the ability for the receiver to detect the presence of energy level for non 802.11 sources on the media for example, the background noise of interference sources of, or non-identifiable 802.11 transmissions that may have been corrupted and cannot be decoded. The detection of energy should sample the media to determine if the energy is always present. In addition, detection of energy requires defining a threshold, which will determine whether the energy level is enough to be able to be reported to the PHY layer, as being occupied. This is usually called the threshold ED or the level of sensitivity of CCA. The ED is generally much lower for valid 802.11 signals that can be decoded using the carrier detect (CS) than it is for non-802.11 signals. There may be a lot of noise and yet not 802.11 carrier. For example, noise ED must be 20 DB above the threshold for an ED signal from a source 802.

4.4.2.3 Carrier Sense (CS)

Carrier sense refers to the ability of the radio receiver to detect an incoming signal and decoding an 802.11 preamble. The role of Carrier Sense differs from CCA, CCA detects whether the media is occupied in the absence of transmission, if simply by noise. Carrier Sense should indicate the occupation of the media by a valid frame transmission for the duration of the frame. This duration is indicated in the PLCP header field that indicates either the number of microseconds or the number of bytes required for the transmission of the MPDU frame payload. In the latter case (number of bytes) this field is then used in combination with the RATE field (identifying the modulation used for the payload) to determine the time required for the transmission MPDU. Length and RATE fields give the receiver the information required to demodulate the frame and determine how long the media will be busy.

4.5 PMD Functions

The PLCP sub layer prepares frames for the radio transmission. PMD sub layer converts the bits into radio wavelengths.
It is interesting to note that while in the 802.11 specifications, PHY is divided into two parts, PLCP and PMD, this sometimes doesn't reflects in implementations. The preamble is then regarded as part of the PPDU, which is described in the PLCP part.
The PMD (Physical media Dependeur) takes care of the encoding/decoding of the data on the radio media. The encoding of the data occurs at three levels:
• Encoding of bits into symbols, a symbol corresponds to a group of bits.
• Encoding of a symbol with timing information.
• Modulation

The PMD with PLCP functions interface uses a state machine that has 3 main states
- Carrier Sense: Gives the state of the media
- Transmit: Sends the bytes of a data packet
- Receive: Receives the bytes of a data packet

4.5.1 PMD Function listing
PMD_RSSI.indication

This primitive, generated by the PMD sub layer, provides the receive signal strength to the PLCP and MAC

```
PMD_RSSI.indication(

RSSI

)
```

The RSSI is a measure of the RF energy received by the HT PHY. This primitive is generated by the PMD to the PLCP after the reception of the HT training fields.

PMD_RCPI.indication

This primitive, generated by the PMD sub layer, provides the received channel power indicator to the PLCP and MAC.

```
PMD_RCPI.indication(

RCPI

)
```

The RCPI is a measure of the channel power received by the OFDM PHY. RCPI is measured by the PMD when the OFDM PHY is in the receive state. It is generated at the end of the last received symbol and sent to the PLCP with this primitive.

PMD_TXPWRLVL.request

This primitive, generated by the PHY PLCP sub layer, selects the power level used by the PHY for transmission.

```
PMD_TXPWRLVL.request(

TXPWR_LEVEL

)
```

TXPWR_LEVEL selects which of the transmit power levels should be used for the current packet transmission. The number of available power levels shall be determined by the MIB parameter aNumberSupportedPowerLevels. This primitive shall be generated by the PLCP sub layer to select a specific transmit power.

PMD_TXPWRLVL immediately sets the transmit power level to the level given by TXPWR_LEVEL.

PMD_TX_PARAMETERS.request

This primitive, generated by the PHY PLCP sub layer, selects the related parameters used by the PHY for transmission.

```
PMD_TX_PARAMETERS.request(
MCS,
```

```
CH_BANDWIDTH,

CH_OFFSET,

STBC,

GI_TYPE,

ANTENNA_SET,

FEC_CODING,

PMD_EXPANSIONS_MAT,

PMD_EXPANSIONS_MAT_TYPE

)
```

This primitive shall be generated by the PLCP sub layer to select a specific transmit parameter. Upon receipt of PMD_TX_PARAMETERS, the PMD immediately sets the transmit parameters for all subsequent PPDU transmissions.

PMD_CBW_OFFSET.indication

This primitive, generated by the PMD sub layer, provides the bandwidth and channel offset of the received frame to the PLCP and MAC.

```
PMD_CBW_OFFSET.indication(

CH_BANDWIDTH,

CH_OFFSET

)
```

CH_BANDWIDTH represents channel width in which the data are transmitted and the transmission format. This primitive is generated by the PMD when the OFDM PHY is in the receive state. It shall be available continuously to the PLCP that, in turn, shall provide the parameter to the MAC sub layer as part of the RXVECTOR.

PMD_CHAN_MAT.indication

This primitive, generated by the PMD sub layer, provides the channel response matrices to the PLCP and MAC.

```
PMD_CHAN_MAT.indication(

CHAN_MAT

)
```

The CHAN_MAT parameter contains the channel response matrices that were measured during the reception of the current frame.

It is available continuously to the PLCP.

The PLCP sub layer passes the data to the MAC sub layer as part of the RXVECTOR.

PMD_TXSTART.request

This primitive, generated by the PHY PLCP sub layer, initiates PPDU transmission by the PMD layer. It has no parameters.

It is generated by the PLCP sub layer to initiate the PMD layer transmission of the PPDU.

The PHY-TXSTART.request primitive is provided to the PLCP sub layer prior to issuing the PMD_TXSTART command.

PMD_DATA.request

```
PMD_DATA.request(

TXD_UNIT

)
```

The TXD_UNIT parameter represents a single block of data that, in turn, shall be used by the PHY to be encoded into an OFDM transmitted symbol.

This primitive is generated by the PLCP sub layer to request transmission of one OFDM symbol.

PMD_DATA.indication

This primitive defines the transfer of data from the PMD entity to the PLCP sub layer.

```
PMD_DATA.indication(

RXD_UNIT

)
```

The RXD_UNIT parameter represents either a SIGNAL field bit or a data field bit. This primitive, generated by the PMD entity, forwards received data to the PLCP sub layer. The PLCP sub layer decodes the bits and either interprets them as part of its own signaling or passes them to the MAC sub layer as part of the PSDU.

PMD_TXEND.request

This primitive, generated by the PHY PLCP sub layer, ends PPDU transmission by the PMD layer.

This primitive has no parameters. It is generated by the PLCP sub layer to terminate the transmission of a PPDU by the PMD sub layer.

PMD_TXEND.confirm

This primitive, generated by the PMD entity, indicates the end of PPDU transmission by the PMD layer. It is generated at the 4 μs boundary following the trailing boundary of the last symbol transmitted.

This primitive has no parameters. It is generated by the PMD entity to inform the PLCP sub layer that transmission of the last symbol of the PPDU is complete. This completion is used as a timing reference in the PLCP state machines.

PMD_DATA.request

This primitive defines the transfer of data from the PLCP sub layer to the PMD entity.

The primitive shall provide the following parameter:

```
PMD_DATA.request(

TXD_UNIT

)
```

The TXD_UNIT parameter takes on the value of either 1 or 0 for DBPSK modulation or the dibit combination 00, 01, 11, or 10 for DQPSK modulation. This parameter represents a single block of data, which, in turn, is used by the PHY to be differentially encoded into a DBPSK or DQPSK transmitted symbol. The symbol itself is spread by the PN code prior to transmission.

This primitive is generated by the PLCP sub layer to request transmission of a symbol. The data clock for this primitive is supplied by the PMD layer based on the PN code repetition.

The PMD performs the differential encoding, PN code modulation, and transmission of the data.

PMD_DATA.indication

This primitive defines the transfer of data from the PMD entity to the PLCP sub layer.

The primitive shall provide the following parameter:

```
PMD_DATA.indication(

RXD_UNIT

)
```

The RXD_UNIT parameter takes on the value of 1 or 0 for DBPSK modulation or as the dibit 00, 01, 11, or 10 for DQPSK modulation. This parameter represents a single symbol that has been demodulated by the PMD entity.

This primitive, which is generated by the PMD entity, forwards received data to the PLCP sub layer. The data clock for this primitive is supplied by the PMD layer based on the PN code repetition.

The PLCP sub layer either interprets the bit or bits that are recovered as part of the PLCP or passes the data to the MAC sub layer as part of the MPDU.

PMD_TXSTART.request

This primitive, which is generated by the PHY PLCP sub layer, initiates PPDU transmission by the PMD layer.

The semantics of this primitive are as follows:

PMD_TXSTART.request

This primitive is generated by the PLCP sub layer to initiate the PMD layer transmission of the PPDU. The PHY-DATA.request primitive is provided to the PLCP sub layer prior to issuing the PMD_TXSTART command. PMD_TXSTART initiates transmission of a PPDU by the PMD sub layer.

PMD_TXEND.request

This primitive, which is generated by the PHY PLCP sub layer, ends PPDU transmission by the PMD layer. The semantics of the primitive are:

PMD_TXEND.request

This primitive is generated by the PLCP sub layer to terminate the PMD layer transmission of the PPDU. PMD_TXEND terminates transmission of a PPDU by the PMD sub layer.

PMD_ANTSEL.request

This primitive, which is generated by the PHY PLCP sub layer, selects the antenna used by the PHY for transmission or reception (when diversity is disabled).

The primitive shall provide the following parameter:

```
PMD_ANTSEL.request(

ANT_STATE

)
```

ANT_STATE selects which of the available antennas should be used for transmit. The number of available antennas is determined from the MIB table parameters aSuprtRxAntennas and aSuprtTxAntennas.

This primitive is generated by the PLCP sub layer to select a specific antenna for transmission or reception (when diversity is disabled).

PMD_ANTSEL immediately selects the antenna specified by ANT_STATE.

PMD_ANTSEL.indication

This primitive, which is generated by the PHY PLCP sub layer, reports the antenna used by the PHY for reception of the most recent packet.

The primitive shall provide the following parameter:

```
PMD_ANTSEL.indication(

ANT_STATE

)
```

ANT_STATE reports which of the available antennas was used for reception of the most recent packet.

This primitive is generated by the PLCP sub layer to report the antenna used for the most recent packet reception. PMD_ANTSEL immediately reports the antenna specified by ANT_STATE.

PMD_RATE.request

This primitive, which is generated by the PHY PLCP sub layer, selects the modulation rate that is used by the DSSS PHY for transmission.

PMD_RATE.indication

This primitive, which is generated by the PMD sub layer, indicates which modulation rate was used to receive the MPDU portion of the PPDU. The modulation is indicated in the PSF.

PMD_SQ.indication

This optional primitive, which is generated by the PMD sub layer, provides to the PLCP and MAC entity the SQ of the DSSS PHY PN code correlation.

PMD_CS.indication

This primitive, which is generated by the PMD, shall indicate to the PLCP layer that the receiver has acquired (locked) the PN code (DSSS) and data are being demodulated.

PMD_ED.indication

This optional primitive, which is generated by the PMD, shall indicate to the PLCP layer that the receiver has detected RF energy indicated by the PMD_RSSI primitive that is above a predefined threshold.

The PMD_ED (energy detect) primitive, along with the PMD_SQ, provides CCA status at the PLCP layer through the PHY-CCA primitive. PMD_ED indicates a binary status of ENABLED or DISABLED.

PMD_ED is ENABLED when the RSSI indicated in PMD_RSSI is greater than the

ED_THRESHOLD parameter. PMD_ED is DISABLED when the PMD_RSSI falls below the energy detect threshold.

This primitive is generated by the PHY when the PHY is receiving RF energy from any source that exceeds the ED_THRESHOLD parameter.

This indicator is provided to the PLCP for forwarding to the MAC entity for information purposes through the PMD_ED indicator.

PMD_ED.request

This optional primitive, which is generated by the PHY PLCP, sets the energy detect ED_THRESHOLD value.

The primitive shall provide the following parameter:

```
PMD_ED.request(

ED_THRESHOLD

)
```

ED_THRESHOLD is the value that the RSSI indicated shall exceed for PMD_ED to be enabled.

The receipt of PMD_ED immediately changes the ED threshold as set by the ED_THRESHOLD parameter.

PHY-CCA.indication

This primitive, which is generated by the PMD, indicates to the PLCP layer that the receiver has detected RF energy that adheres to the CCA algorithm.

The PHY-CCA primitive provides CCA status at the PLCP layer to the MAC.

This primitive is generated by the PHY when the PHY is receiving RF energy from any source that exceeds the ED_THRESHOLD parameter (PMD_ED is active).

This indicator is provided to the PLCP for forwarding to the MAC entity for information purposes through the PHY-CCA indicator. This parameter indicates that the RF medium may be busy with an RF energy source.

4.5.2 GUARD INTERVAL

It is the time between two OFDM symbols.

In OFDM to parse a symbol, one must first differentiate a symbol from the following symbol and therefore realize that the incoming flow conveys a new symbol. Then it is time to recognize this new symbol as a specific code. In 802.11 it should take 3.6μsec to parse a symbol.

The "Guard Interval" is worth four to eight times the maximum delay between two reflections of a signal time. This time varies between 100 and 150ns, and is higher in open terrain. The "Guard Interval" is longer than the longest PHY delay.

The guard interval is the time between two symbols (the smallest unit of data sent at a time) consecutive. The guard interval is necessary to compensate for the multi-path effects that otherwise could cause inter-interference (ISI) symbols.

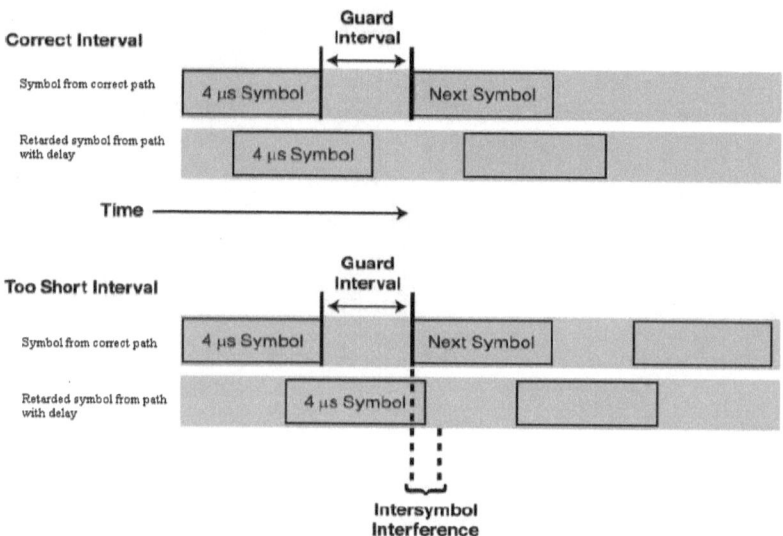

Figure 66 Guard interval

It is the equivalent of the pause between the words that a speaker does when she wants to compensate for an echo.

When a station emitting the frame transmits to another station, the wireless signal bounces off walls, obstacles etc., and traverses multiple paths to the station receiving the frame. This phenomenon, known as the multi-path effect, is a common distortion in wireless channels such as 802.11. As a result of the multi-path effect, different copies of the same signal arrive at the station receiving the frame. These different copies of the same signal are delayed with respect to each other. This means that the energy from one symbol bleeds into the next symbol, and corrupts its signal.

One key principle of OFDM is that since low symbol rate modulation schemes and since the duration of each symbol is long, it is feasible to insert a guard interval between the OFDM symbols, thus eliminating the intersymbol interference.

OFDM symbols have a guard interval between them. In a typical network, the value of the guard interval is chosen to be a little larger than the maximum multipath delay spread of the network, i.e., the maximum delay difference between delayed copies of the signal.

An OFDM station in order to decode, converts the received symbol to a frequency representation by taking an FFT of the symbol. In order to do so while ensuring that the symbol is not corrupted by multipath noise from the previous symbol, the receiving station the frame should use only the parts of the OFDM symbol that are considered as safe from multipath reflections. In a typical network, the guard interval has a small amount of slack to allow for packet detection errors. This means that the station receiving the frame has a corresponding amount of slack in the choice of where to align the station receiving the frame FFT window in a symbol.

The cyclic prefix, which is transmitted during the guard interval, consists of the end of the OFDM symbol copied into the guard interval, and the guard interval is transmitted followed by the OFDM symbol. The reason that the guard interval consists of a copy of the end of the OFDM symbol is so that the receiver will integrate over an integer number of sinusoid cycles for each of the multipaths when it performs OFDM demodulation with the FFT.

The normal 802.11 stations guard interval has a duration of 800 nano seconds, but 802.11n devices have the ability to reduce this guard interval to only 400 nano seconds. Shorter guard intervals would lead to more interference and reduced throughput, while a longer guard interval would lead to a downtime unwanted on the radio channel. A short guard interval (SGI) increases the maximum rate of data transfer by 11 per cent while providing a separation between symbols sufficient for most indoor environments. The short guard interval is not suitable for open systems environments.

4.5.3 MODULATIONS
4.5.3.1 OFDM modulation.
OFDM is a method of modulation and encoding of information on multiple carrier frequencies. If a sub-carrier experiences fading at some point that sub-carrier can be amplified without affecting other sub-carriers, or less data could be sent over it. OFDM has been developed for broadband digital communication.

A large number of closely spaced signals orthogonal sub-carriers are used to carry data. The data are divided into several parallel data streams named channels, one for each sub-carrier. Each sub-carrier is modulated with a conventional modulation at a quite slow symbol rate scheme. For increasing the Spectral Efficiency of the channel it is highly desirable that the sub-carriers are non-overlapping with each other. It could be done by introducing a guard band between each pair of sub-carriers which will spread them out over the channel. The downside of this is that a significant amount of bandwidth will be unused to serve as separators.

In OFDM, each sub-carriers are transmitted in mutually orthogonal frequencies, meaning that the extremum of one sub-carrier will coincide with the null of an adjacent sub-carrier. Because of this, there will be no ISI between them which diminishes the need for a guard band. This way a larger part of the band can be used for transmitting data. The total data rate is the sum of all sub-carriers data rate and similar to that of conventional modulation schemes of a single carrier in the same bandwidth.
Modulating processes are said orthogonal when an ideal sub-carrier receiver can completely reject arbitrarily strong unwanted signals.

The main advantage of the OFDM over single-carrier modulation resides in its ability to cope with harsh conditions without complex equalization filters. The Channel equalization is simplified because OFDM may be regarded as using many narrow band signals slowly modulated rather than a single broadband signal modulated quickly.

In telecommunications, the equalizers are used to make the frequency response flat to one end of the spectrum to the other. When a channel is "equalized" the signal attributes at the input, are accurately reproduced at the output. Equalization output is also used to better identify binaries symbols in the iterative decoder, which is often LDPC.

To better achieve equalization some OFDM sub-carriers are not used for transmitting the signal. These are the null sub-carriers which are used for DC sub-carrier(s) or guard sub-carriers. The DC sub-carrier bear "0" as value in order to reduce problems from analog/digital converters and carrier feed-through.

This is somewhat similar in goal to the pilot tones that are inserted in VHT-LTF.

The low symbol rate makes use of a guard interval between symbols possible (guard intervals are used to ensure that distinct OFDM symbols do not interfere with each other). A guard interval can eliminate inter-symbol interference (ISI is a form of distortion of a signal in which one symbol interferes with the following symbols).

4.5.3.2 Modulation Constellation

The IEEE 802.11ac design philosophy closely follows that of the IEEE 802.11n standard. Most of the mandatory features in 802.11n are preserved. The transmission on multiple spatial streams is not required in 802.11ac but the 80 MHz bandwidth channel support is required. One of the reasons for this requirement is that the multiple spatial streams support involves having at least the same number of antennas and channels reception which would add significant costs. As a result, a large number of 802.11n or 802.11ac devices available on the market can support only a single spatial stream. This is particularly the case of smartphones, which due to their small size cannot host two 5 GHz antennas, a GPS antenna and one or several cellular radio antennas.

The channel at 80 MHz is indeed composed of 234 sub carriers for data, on a total of 256 sub-carriers involved in a fast Fourier transform (FFT). That's more than twice the number of sub-carriers of data found in the channel at 40 MHz, which it has only 108 sub-carrier data.

In addition, 802.11ac increases the size of the modulation QAM of 64 to 256 QAM constellations. Patterns of modulation / coding (MCS) are limited to identical MCS for each stream (not unequal modulation as in 802.11n).

OFDM has natively a high PAPR (Peak to Average Power Ratio) which reduces the efficiency of power amplifiers. To mitigate this effect, the sub-carriers of the upper 20 MHz sub-bands are rotated.

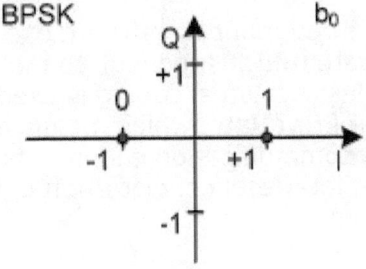

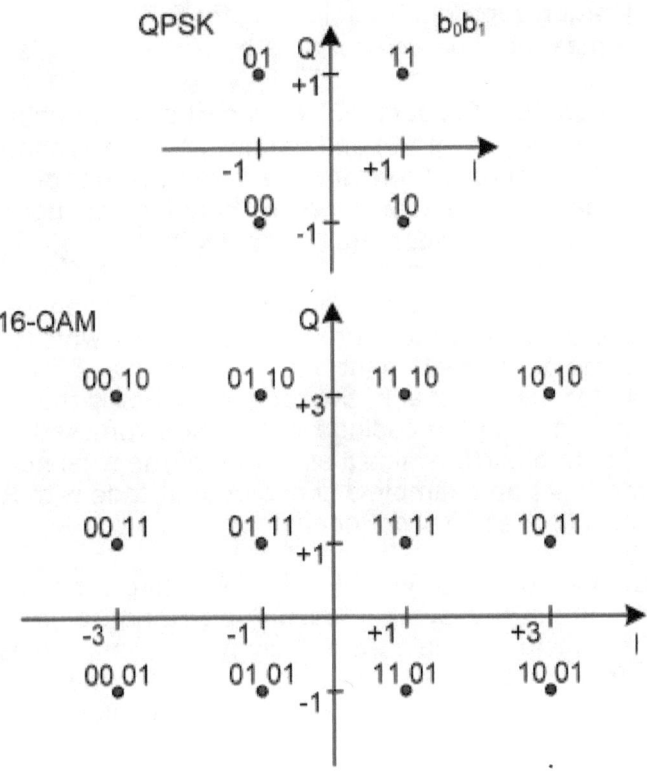

Figure 67 Examples of modulation

4.5.3.3 BCC coding

OFDM enables high throughput by efficient use of the spectrum but often the error rate (BER) is too high to make it really useful. In most wireless systems, coding is used because it enables excellent Bit Error Rate, which means a low number of received bits over a communication channel that have been altered due to noise, interference, distortion or bit synchronization errors.

In 802.11ac there are two codes, one is BCC which support is mandatory, and the other is LDPC which is optional.

The specification supports the following for BCC:
- The maximum delivery of data by encoder BCC is 600 Mbps
- The number of coders BCC for a particular combination of MCS, Nsts (# of SS) and bandwidth is determined by the short IM data flows and the same number of encoders are used for the corresponding flow normal GI
- The number of coders BCC is not limited

BCC code, works by adding redundant bits. It is not an iterative code. After the BCC encoder, it exists a puncturer which removes some of the redundant bits. The parameter that is responsible for the removal of the redundant bits is the coding rate. The definition of the coding rate can be expressed, "as the number of data bits transmitted as a ratio of the total number of coded bits". As an example a convolutional code with R=3/4 has only 25% of thee bits redundant

A convolutional code is a type of error-correcting code in which each m-bit information symbol (each m-bit string) to be encoded is transformed into an n-bit symbol, where m/n is the code rate (n ≥ m) and the transformation is a function of the last k information symbols, where k is the constraint length of the code.

4.5.3.4 LDPC coding
It happens that most of those errors are related to a bad equalization of sub-carriers, so error correcting codes that can use information from the equalization are of great help. Unfortunately those codes can't produce a good result in a

straightforward manner, they have to make hypothesis, test them and select the best one: They are iterative codes. There are several candidates such as Turbo-codes or LDPC. LDPC is an option in 802.11ac.

Low Density Parity Check (LDPC) code is a linear error correcting code, a method of transmitting a message over a noisy transmission channel. LDPC codes are capacity-approaching codes, which means that practical constructions exist that allow the noise threshold to be set very close to the theoretical maximum (the Shannon limit). The noise threshold defines an upper bound for the channel noise, up to which the probability of lost information can be made as small as desired. Using iterative belief propagation techniques, LDPC codes can be decoded in time linear to their block length. The absence of encumbering software patents has made LDPC attractive to some. LDPC codes are also known as Gallager codes, in honor of Robert G. Gallager.

As with other codes, optimally decoding an LDPC code on the binary symmetric channel is an NP-complete problem (a problem impossible to solve in a finite time), although techniques based on iterative belief propagation used in practice lead to good approximations. In contrast, belief propagation on the binary erasure channel is particularly simple where it consists of iterative constraint satisfaction. Unlike the binary symmetric channel, in binary erasure channel when the receiver gets a bit, it is 100% certain that the bit is correct. The only confusion arises when the bit is erased.

Such decoders use the likelihood of data to reconcile differences between the decoder and some criteria/constraints. The convolutional decoder generates a hypothesis for the pattern of m bits in the payload sub-block. The hypothesis bit-patterns are compared with what the criteria would suggest, and if it differs at the next iteration the decoder improves the likelihoods it has for each bit in the hypotheses. This iterative process continues until the decoder come up with an output that satisfies the criteria/constraints, typically in 15 to 18 cycles.

An analogy can be drawn between this process and that of solving cross-reference puzzles like crossword or Sudoku. Consider a partially completed, possibly garbled crossword puzzle. Two puzzle solvers (decoders) are trying to solve it: one possessing only the "down" clues (parity bits), and the other possessing only the "across" clues. To start, both solvers guess the answers (hypotheses) to their own clues, noting down how confident they are in each letter (payload bit). Then, they compare notes, by exchanging answers and confidence ratings with each other, noticing where and how they differ. Based on this new knowledge, they both come up with updated answers and confidence ratings, repeating the whole process until they converge to the same solution.

As a more precise example, consider that the valid codeword, 101011, is transmitted across a noisy channel (actually it works only with binary erasure channels) and received with the first and fourth bit erased to yield ?01?11. The two bits cannot be recovered simultaneously, because all of the constraints connected to it have more than one unknown bit. In order to proceed with decoding the message, constraints connecting to only one of the erased bits must be identified. If one constraint is satisfied if the fourth bit have been 0, then this procedure is iterated. The new value for the fourth bit can now be used in conjunction with the criteria/constraint set to recover the first bit.

4.5.3.5 Pilot, Null sub carriers and channelization

Pilot tones are used for phase and frequency tracking and training. The 'pilot tones' are sub OFDM carriers which have a width of 345 KHz. They convey symbols known to both the transmitter and the receiver.

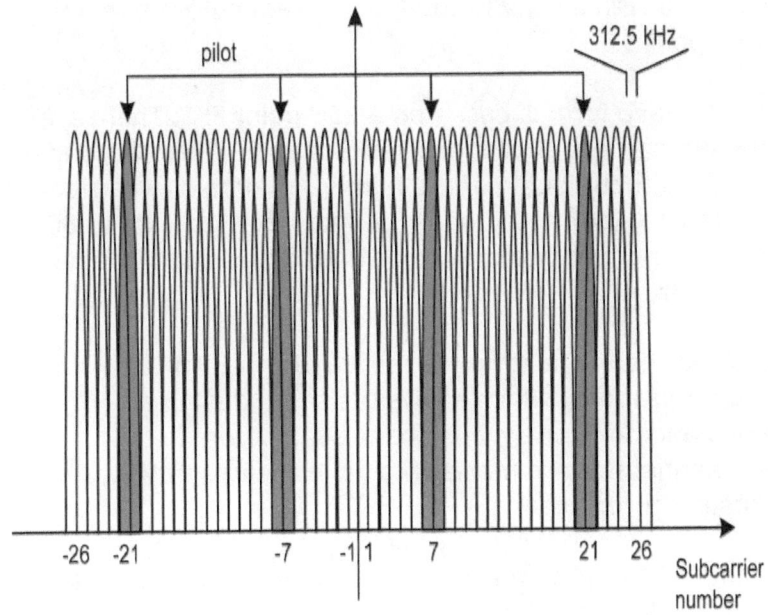

Figure 68 OFDM Pilot sub carriers in 802.11a

The transmission uses one or two 20 MHz channel and each channel is divided into 64 sub carriers. Four pilot signals are inserted in sub carriers-21, -7, 7 and 21. Pilot signals are symbols known in advance, to avoid various problems depending on sub-carrier position in the channel such as variation in gain. A different cyclic shift is applied to each of these pilot signals.

There are also 14 Null tones: {-128, ... -123, -1, 0, 1, 123, ..., 127}. To better achieve equalization Null tones are the null sub-carriers which are used for DC sub-carrier(s) or guard sub-carriers. The DC sub-carrier bear "0" as value in order to reduce problems from analog/digital converters and carrier feed-through.

To lower the PAPR, for a 80 MHz channel, the pilot sequence is designed such that a rotation [1-1-1-1] is done on the Sub 20

MHz channels, with a copy to the Nsts stream before streaming CSDs are applied.

The 80 MHz wave form is based on a 256-point FFT. There are 234 data sub-carriers, 8 pilot sub carriers and 14 carriers under zero. It's more than double of the number of data sub-carriers of the 40 MHz waveform (108 tones for data), so that data at 80 MHz, have a throughput of the more than double than the data rate at 40 MHz.

The 160 MHz channel is simply a replica of two 80MHz channels without optimizations, which allows the same design at contiguous and non-contiguous channels. In addition for 160MHz channel there is no need to have the two 80MHz oscillators synchronized.

The number of OFDM symbols in the data field is calculated using the field length in L-SIG.

Both for BCC and LDPC, all bits (including MAC and PHY pad bits) are encoded. When the BCC encoding is used, the data field consists of the SERVICE field on 16 bits, the PSDU pad bits, the PHY pad bits and end sequence of bits (6NES bit), in that order.
When the LDPC coding is used, the data field consists of the SERVICE field on 16 bits and the PSDU pad bits. No bit of filling PHY is included when the LDPC coding is used.

Bandwidth (MHz)	Number of Sub-carriers	Sub-carriers
20	64	-28 to -1 and «1» to 28
40	128	-58 to -2 and 2 to 58
80	256	-14.1.to -2 and 2 to 122
160	512	-250 to-130, -126 to -6, 6 to 126 and 130 to 250
80+80	256 per 80MHz Channel	-14.1.to -2 and 2 to 122

Figure 69 Sub carriers and channelization

4.5.3.6 Modulation and coding scheme (MCS)

The specification includes 256 QAM. Support for 256 QAM by an 802.11ac station is optional.

The specification maintains the 802.11n modulation, interlacing, and coding the architecture.

The specification minimizes the number of additional options MCS.

The specification includes support for a different MCS for each station in a MU-MIMO transmission DL.
The specification includes an extended MCS for additional space flows supported.
The MCS mono user are presented below. MCS 9 will not be used in 20 MHz bandwidth broadcasts.

MCS	Modulation	Coding Rate
0	BPSK	½
1	QPSK	½
2	QPSK	¾
3	16-QAM	½
4	16-QAM	¾
5	64-QAM	2/3
6	64-QAM	¾
7	64-QAM	5/6
8	256-QAM	¾
9	256-QAM	5/6

Figure 70 802.11ac mono user MCSs

For the BCC, some of the combinations of MCS - coding NSS are excluded from the MCS table to avoid padding additional symbols. Authorized MCS are selected such that the number of bits encoded in each OFDM symbol contains a whole number of blocks punched from all measurement systems.

4.5.3.7 Transmit center frequency
The carrier (LO) and the clock frequencies of symbol for all frequency segments and broadcast channels, is derived from the same reference oscillator.

It is not required that the phase of the carrier frequency is correlated with the portions of frequency lower and upper to 80 MHz signal transmitted to PPDU to 160 MHz.

4.5.3.8 Transmit spectral mask

The transmission spectrum mask for channels at 20 MHz, 40 MHz, 80 MHz and 160 MHz PPDU is shown below.

throughput (MHz)	0 dBr	-20 dBr	-28 dBr	-40 dBr	Maximum Transmit OOB Emission
20	+/- 9	+/- 11	+/- 20	+/- 30	-53 dBm/MHz, if transmit power is below 0dBm
40	+/- 19	+/- 21	+/- 40	+/- 60	-56 dBm/MHz, if transmit power is below 0dBm
80	+/- 39	+/- 41	+/- 80	+/- 120	-59 dBm/MHz, if transmit power is below 0dBm
160	+/- 79	+/- 81	+/- 160	+/- 240	-59dBm/MHz, if transmit power is below 3dBm

Figure 71 Transmit spectral mask for various throughput PPDUs

In case of non-contiguous to 160 MHz PPDU emission mask is constructed as follows:
Place two masks transmission (TX) at 80 MHz, one for each segment
v1 = value of mask a mask to 80 MHz
v2 = value of mask on the other mask 80 MHz
For frequencies in the range to (-40 dBr < v1 < dBr-20) and (-40 dBr < v2 < dBr-20)
Mask value = v1 + v2 (sum in the linear field)
For frequencies which are not in the range {(-20 dBr <v1 <0 dBr) or (-20 dBr < v2 < 0 dBr)}
Mask value = max (v1, v2)

271

For all other frequencies
Mask value = linear interpolation (in dB field)

4.5.3.9 Spectral flatness

The average energy in the PPDU sub-carriers can deviate more than that shown in the table below.

throughput	Average variation	
	+4/-4 dB	+4/-6 dB
20	+/- 1-16	Outer sub carriers
40	+/- 2-42	Outer sub carriers
80	+/- 2-84	Outer sub carriers
160	-	All sub carriers

Figure 72 Average energy variation on sub carriers

A non-contiguous channel at 160 MHz can transmit a contiguous waveform at 160 MHz by placing its two 80 MHz segments adjacent to each other.
It would be difficult for these transmissions to meet the requirement of + 4 dB /-4 near the center of the contiguous zone to 160 MHz wave form.

4.5.3.10 Modulation accuracy

Transmission requirements EVM to 256 - QAM ¾ and 256 - QAM 5/6 is-30 dB and -32 dB respectively. The other transmission (TX) EVM for MCS requirements are as in the 802.11n standard.

4.5.4.1 Introduction

Physically, the most common format for a network radio device is a "system on chip" for a Smartphone or a dedicated chipset on a PC. Possibly it may be a USB adapter for Ethernet 802.11 card.

Electrically, the 802.11 card is divided into two main sections: the analog tuner (PHY layer) and the digital section (MAC, or Medium Access Control)

In an access point, in addition to the radio equipment and digital, a circuit is used to interface with a wired local area network (if business) or ADSL (domestic cases) network.

Description of the main hardware blocks:

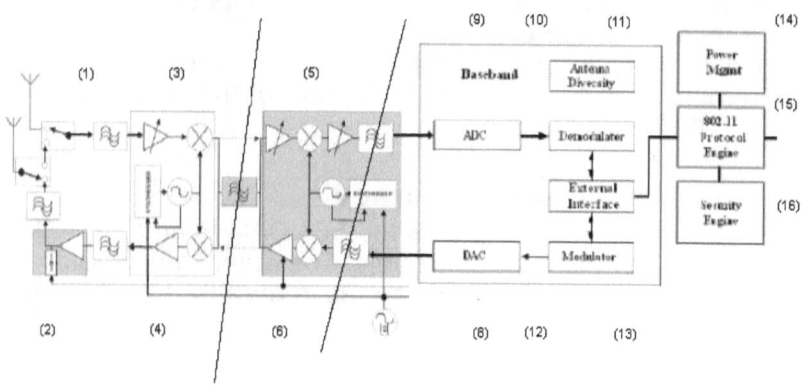

Figure 73 Emitter main hardware blocks

We see that there are three main blocks: The high frequency radio part at left, the intermediate frequency part in the middle and the MAC layer to the right. Two blocks to the left but also the digital and analog converters as well as the modems belong to the physical layer and PHY/PLCP specifically.

274

(1) It is the entry point of the radio signals in the receiver. A switch allows choosing the best antenna to take advantage of the spatial diversity. Just after this switch you can see a transformer to electrically isolate the electronic circuits from of the receiver antenna.

(3) This receiver is equipped with an automatic gain control (AGC). A frequency mixer will transpose the signal which is at very high frequency into a constant intermediate frequency and for which detection circuits are profiled. Below this mixer you can see the frequency generator which provides the signal to be mixed with the received signal.

(2) It is the output of the high frequency amplifier. A switch allows choosing between the 'emission' and 'reception ' mode. A high frequency amplifier powers the antenna. This amplifier is isolated from the rest of the electronics by a transformer.

Among the two design of radio amplifier, switching amplifiers (class C amplifiers) cost less and use less power than linear amplifiers of the same power output. However, they only work with a relatively constant amplitude signals such as modulation (FSK or PSK) modulation and CDMA, but not with QAM and OFDM.

(4) The frequency generator generates a signal that is mixed with the signal from intermediate frequency block emerging from the transformer separating the part of the intermediate frequency electronic circuits from the high frequency circuits. The result is a high-frequency signal modulated with the information to emit, which is first pre-amplified before being sent to the power amplifier via transformer insulating the high frequency block.

(5) The received signal is transferred to a mixer which will have the effect of transpose on a lower intermediate frequency. This new range of frequency is low enough for making it possible for ordinary circuits to operate. We will see below that this is the output of a modem.

(6) It is the reverse of what is achieved in (5). Here we take a signal from the modem via an isolation transformer and we modulate this signal in the band allocated to the intermediate

frequency. Once this is done the modulated signal is transferred to the high frequency radio block.

(7) Nc

(8) and (9) it is a modem, it means an electronic circuit that is designed to process digital signals into low frequency signals, and the other way round. These signals can be modulated with OFDM or CCK for example. If the waveform is complex there is need of an analog-to-digital converter.

(10) Antenna diversity: the goal here is to take advantage of having several radio (the two major parties at left in the figure below) systems. There are several ways to do that, it will be described in detail later. First way is to search what is the radio system that has the best signal-to-noise ratio at a given moment, another method is to add the signals, and initially the noise is not added because the noise in two radio systems (and the roads traveled radio) is not synchronized. So the result has a better overall signal-to-noise ratio.

Figure 74: Receiver main hardware blocks

(11) and (12) Those are modems. As seen above, the goal is here to convert digital information in the form of radio signals

essentially by the means of a combination of changes of phase, amplitude and frequency. These modulations can be quite complex so as to introduce additional features, like the ability to easily retrieve the clock (which is very useful in case of phase modulation).

(13) Nc

(14) Power management: it is a layer of software for the management of energy based on a predefined policy.

(15) Nc

(16) Security engine: it's a software layer responsible for the association of a station to an access point by means of secret keys as well as of the data encryption.

(17) Nc

4.5.4.2 Emitter Conception

The unit of data (PSDU) PHY service is handled by the PHY transmission chain to form a physical layer protocol data unit (PPDU). The PSDU bits are first filled with all the OFDM symbols, and scrambled. The resulting bits are then encoded using either a Binary Convolutional Coding (BCC) or a Low Density Parity Control (LDPC). When BCC is used, there may be '1' to 12 terms of encoders of error correction (FEC), based on the maximum data rate. The coded bits are rearranged in up to 8 space flow by flow Analyzer. Each block of data bits is interlaced and a bit-to-symbol lookup table is carried out for CCB, while the lookup table is performed before "1" QAM tones for LDPC correspondence. When the block of space-time coding (STBC) is used, the STBC block has twice as many outputs as inputs. Results in spatio-temporal release of flows (STS), in which different cyclic shift diversity (SDR) are applied to prevent the movements of beam forming shall not exceed 8.

When the spatial correspondence table is used, there may be strings convey more space-time flow. The frequency field symbols are transposed to symbols in the time field by inverse FFT (IFFT), preceded by a cyclic prefix guard interval (GI), and windowed, before radio-frequency treatment starts and the PPDU is sent on the radio media.

When channels at 160 MHz are used, some of the blocks are duplicated. For the contiguous channels to 160 MHz, the table striping / tone and constellations correspondence functions are duplicated (a block of space flow per 80 MHz channel). For the non-contiguous to 160 MHz, STBC, CSD, and spatial correspondence table functions are included in the duplication.

4.5.4.3 Emitter and MIMO

When the access point uses MU-MIMO to send independent PSDU to different customers, should be treated relevant data in a parallel way. Indeed, in single-mode MIMO (single user MIMO), each destination has access to all signals received at their antennae and uses that information in the decoding process. When it comes to MU-MIMO, there is no cooperation between stations in the receiving process.

The total number of spatial streams must not exceed the maximum of 8. In addition, for channels to 160 MHz and 80 + 80 MHz, each branch processing at 80 MHz being duplicated PSDU

4.5.4.4 Receiver Conception

A successful demodulation is determined by package (PER) less 10% error rate. For 802.11ac the minimum input level depends on the modulation rate coding and bandwidth, as shown in the table below. The 11ac packets used for this test must be 4096 bytes, they use an 800 ns guard interval, BCC and a non-STBC.

	Coding rate	20 MHz	40 MHz	80 MHz	160 or 80+80 MHz
BPSK	1/2	-82	-79	-76	-73
QPSK	1/2	-79	-76	-73	-70
QPSK	3/4	-77	-74	-71	-68
16-QAM	1/2	-74	-71	-68	-65
16-QAM	3/4	-70	-67	-64	-61
64-QAM	2/3	-66	-63	-60	-57
64-QAM	3/4	-65	-62	-59	-56
64-QAM	5/6	-64	-61	-58	-55
256-QAM	3/4	-59	-56	-53	-50
256-QAM	5/6	-57	-54	-51	-48

Figure 75 Minimum input level depends on MCS

4.5.4.5 Receiver minimum input sensitivity

The minimum sensitivity of the receiver, as shown below.

Modulation	Rate (R)	Minimum sensitivity (20 MHz PPDU) (dBm)	Minimum sensitivity (a 40 MHz PPDU) (dBm)	Minimum sensitivity (a 80 MHz PPDU) (dBm)	Minimum sensitivity (a 160 MHz or 80+a 80 MHz PPDU) (dBm)
BPSK	1/2	-82	-79	-76	-73
QPSK	1/2	-79	-76	-73	-70
QPSK	3/4	-77	-74	-71	-68
16-QAM	1/2	-74	-71	-68	-65
16-QAM	3/4	-70	-67	-64	-61
64-QAM	2/3	-66	-63	-60	-57
64-QAM	3/4	-65	-62	-59	-56
64-QAM	5/6	-64	-61	-58	-55
256-QAM	3/4	-59	-56	-53	-50
256-QAM	5/6	-57	-54	-51	-48

Figure 76 Receiver minimum input sensitivity

4.5.4.6 Adjacent channel rejection

The test of adjacent / non-adjacent is defined as follows:
• The power of the desired signal is 3 DB above the level of sensitivity in receive (RX)
• If the desired signal bandwidth sensitivity = 20/40/80/160 MHz
   ◦ The signal interfering (same bandwidth as the signal desired) placed far from the Central frequency of the desired signal
   ◦ the adjacent channel rejection: W = the signal bandwidth desired
   ◦ Discharge channel + non-adjacent: W ≥ 2 * the desired signal bandwidth
• If the desired signal bandwidth is 80 + 80 MHz
   ◦ 80 MHz bandwidth interference signal placed W MHz of the Centre frequency of one of the segments of desired frequency signal
      ▪ the adjacent channel rejection: W = 80 MHz
      ▪ non-adjacent channel rejection: W ≥ 160 MHz
• Increase the power of the interfering signal up to 10% achieved by the minimum required level of rejection for the channel adjacent and non-adjacent as shown below.

Modulation	Rate (R)	Adjacent channel rejection (dB)		Nonadjacent channel rejection (dB)	
		20/40/80/a 160 MHz Channel	80+a 80 MHz Channel	20/40/80/a 160 MHz Channel	80+a 80 MHz Channel
BPSK	1/2	16	13	32	29
QPSK	1/2	13	10	29	26
QPSK	3/4	11	8	27	24
16-QAM	1/2	8	5	24	21
16-QAM	3/4	4	1	20	17
64-QAM	2/3	0	-3	16	13
64-QAM	3/4	-1	-4	15	12
64-QAM	5/6	-2	-5	14	11
256-QAM	3/4	-7	-10	9	6
256-QAM	5/6	-9	-12	7	4

Figure 77 Minimum levels required to reject adjacent and non-adjacent channels

4.6 PMD AND MIMO

4.6.1 MIMO SPECIES

MIMO is a mode where several electromagnetic signals, are sent at the same time, in the same frequency band with various amplitudes and relative phases. MIMO implies the use of several antennas.

A very common experience is that moving an antenna of a FM receiver, can change dramatically the reception condition. Pre-802.11n access points often used two antennas and selected the best in reception depending on the SNR.

As the methods for eliminating multipath (echoes) effects improved it was realized that those methods would help improve SNR by deliberately use several antennas at emission and reception.

MIMO can be classified by the effect used or by the technology enabling it. A MIMO property is the degree of diversity between antennas.

Effects used:

1. Improving SNR by adding the same signal coming on different paths (STBC).

2. Improving throughput by adding throughput on different paths.

3. Beam-forming where constructive interferences are used to achieve directional beams.

The first and second methods are quite close, they differentiate only by the fact that in the first method only one signal is sent (but with a different coding scheme to improve spatial diversity) and the second uses several spatial streams. However the second method works only in case there is an overall very good SNR, and in the first method if a better SNR is achieved then a throughput enhancement is possible. In beam-forming the phase difference at reception results in that if interference occurs, at certain aligned points the received signal will be strengthened, and at others places it will be attenuated. MIMO allows increasing the signal to noise ratio and therefore the throughput.

In order to make MIMO possible a good knowledge of the channel state is important. Even if it's possible to infer the CSI from past receptions, this possibility was removed in 802.11ac who uses only explicit feedback.

It should be noted that MIMO exacerbates some OFDM problems such as non homogeneous SNR along the sub-carriers or PAPR.
Beam-forming is very far from forming perfect beams, actually it improve the SNR for a few dBm for the stations in the right direction, but all stations in the same range as the target are able to receive some energy.

In order to implement MIMO system into 802.11ac, some new function blocks are the Stream Parser (SP) and Spatial Mapper (SM).

A compressed digital source is fed into a simplified transmitting block with the functions of error control coding and mapping to complex modulation symbols. The latter produces several separate symbol streams which are independent.

Each is then mapped onto one of the multiple TX antennas. Mapping may include linear spatial weighting of the antenna elements or linear antenna space-time precoding. After frequency conversion and amplification, the signals are launched into the wireless channel. At the receiver, the signals are captured by possibly multiple antennas and demodulation and demapping operations are performed to recover the message. The level of intelligence, complexity, and a priori channel knowledge used in selecting the coding scheme and antenna can vary depending one real system requirement.

4.6.2 MU-MIMO

One important feature in 802.11ac, is to improve MIMO performance with multiple users. Indeed before the MIMO mode transmission was only possible between the access point and only one station at a time. What was not necessarily a so serious problem as all stations were visible from each other, and therefore were, in any case, forced to wait until the channel becomes free to emit (following the principle of CSMA). However with 802.11ac access points are now able to address sets of stations. This introduces the concept of "multi-user" transmission. MU-MIMO allows having several MIMO simultaneously.

However in 802.11ac only the access point is at the origin of the exchange, this mode is known as DL-MU-MIMO. One practical reason is that to achieve MIMO it takes at least several antennas by simultaneous exchange (space stream). So it enable at most MU-MIMO to four destinations at the same time, and it must be have at least eight antennas at the emitting end point which is a lot even for an access point.

Previously the MIMO mode was possible only between the access point and only one station at a time. What was not necessarily very relevant as all stations were visible to each other, and therefore they were in any event, forced to wait until the channel was free, to emit (following the principle of CSMA). However nowadays with 802.11ac there are access points that are able to address sets of stations. While accessing a different radio media, these groups form a single logical cell because the access point is able to access all of these stations which can thus communicate with each other via the access point. Under these conditions it is possible to interact with several stations in both MIMO access point.

While accessing a different radio media, each group of stations form a single logical cell because the access point is able to access all of these stations which can thus communicate with each other via the access point independently of the other stations of the BSS.

The number of spatial streams can go up to eight, what is twice 802.11n allows.

These eight MU-MIMO spatial streams can be spread over a maximum of four different receiving stations. It means that there might have four simultaneous beams originating from the access point. These streams also included the possibility of acknowledgment and sounding of channel suitable for multiple users operation.

Sending specific data to multiple users involves multiple and independent stream modulations and pre-coding and specific information. These channels also include the possibility of independent acknowledgment and channel sounding suitable for the operation with multiple users.

In the 80 MHz mode, it would be possible to send frames to simultaneously to four receiving stations at a 866,6 MB/s rate per receiving station, if we assume that all receiving stations can receive two spatial streams. This means that the total data rate of 3.46 Gbs at BSS level is four times greater than what it would have been in a BSS without MU-MIMO.

The data throughput by receiving station is also larger, as each transmission to a station can have a duration of about a quarter of the time needed in a non MU access point, there are four time less contention so that the throughput by station is much better it would have been without MU-MIMO. It's an indirect advantage given by the MU-MIMO.

In practice, the throughput of MU-MIMO gain is somewhat reduced by the fact that acknowledgments are always sent sequentially in time. In addition, as the signal-to-noise ratios were different depending on the paths and the receiving stations, it may that a receiving station cannot support the maximum throughput in a packet of MU-MIMO hence slowing all the station in the group. The 802.11ac standard provides mechanisms to recommend a certain modulation and coding (MCS), but its component suppliers are free to determine how they will actually use these mechanisms.

Another challenge will be how to deal with the change of time reference in the channel, as MU-MIMO requires specific knowledge of the characteristics of the channel in order to minimize the inter-user interference. 802.11ac specifies a compression method of explicit feedback of the beams training to enable both MU-MIMO and single-user MIMO. The fact that one method has been specified should help the market adoption ensuring interoperability between the stations. This is in contrast to the 802.11n standard where the presence of multiple feedback options had compromised the widespread adoption of the mechanism of beam forming radio.

4.6.3 NUMBER OF SPATIAL STREAMS

There are always some confusion about the number of available streams in 802.11.

The standard itself limits the number of spatial streams, however Wi-Fi Alliance even limits it further.

The number of spatial streams is variable for each station engaged in an association with an AP: It is the least of two numbers, the number of antennas at the AP and at the station. For example an AP with 8 antennas and a station with one antenna will be able to use only one spatial stream. The only way the couple (station, AP) can benefit from the high number of antennas at the AP is to use beam-forming at the AP.

Its not possible to benefit at the same time of several MIMO modes with the same set of antennas.

Unlike the 802.11n standard , where each station can support up to four spatial stream (SS) the transmission of eight spatial stream is an option in 802.11ac. However the support of multiple spatial streams is not mandatory so devices and in particular smartphones, can be labeled as being conformed to 802.11ac specification even if they have only one antenna!

Even if the maximum number of spatial streams (Nsts) in mono user transmission is eight, the maximum number of streams (Nsts) for a single user in a multi-user (MU) transmission is 4 but the reason is a bit simple, making MU transmission implies using several antennas, which are removed from the pool of available antennas.

The same modulation and the same encoding speed and the type of encoding is used on all streams in mono user transmission.
The same modulation and the same encoding speed and the type of encoding are used in all streams belonging to each user in a multi-user (MU) transmission.

Scenario	Number of spatial streams at station.	Beam-forming	Mu-Mimo
AP with an antenna, a station with an antenna.	One spatial stream.	no	no
AP with two antennas, a station with one antenna.	One spatial stream.	yes	no
AP with four antennas, two stations with one antennas.	One spatial stream.	yes	Yes, with two sets of two antennas each.
AP with four antennas, four stations with an antenna each.	One spatial stream.	Yes, with one spatial stream.	Yes, but limited to two stations simultaneously, with two sets of two antennas each.
AP with eight antennas, four stations with two antennas.	Two spatial streams.	Yes, with one spatial stream.	Yes, with four sets of two antennas each.

4.6.3 BEAM FORMING

4.6.3.1 Beam forming introduction

The OFDM technique has the great merit of turning a wide radio channel into a set of sub-channels very simple to equalize.

The use of the Fourier transform for modulation and demodulation was proposed for the first time by Saltzberg in 1967 and then by Weinstein et al. in 1971. The realization of perfectly orthogonal analog filters being expensive, this system didn't have the expected success.

The fast calculation of Fourier transform algorithm was invented by Cooley and Tukey, both engineers in the center of IBM research in the early 1960s. Multi-carrier modulations have experienced a revival of interest at the beginning of the 1980s, using a fast Fourier transform.

It had, because of its efficiency, a considerable impact on the development of digital signal processing applications because it reduces the complexity of the modulator and thus the terminals power requirements.

A calculation of the discrete Fourier transform is a product of a matrix by a vector carried out recursively.

The frequency representation of the equation (1) gives us:

$Y(f) = H(f) * S(f) + N(f)$

$Y(f)$ represents the Fourier transform of the received signal.
$H(f)$ is the Fourier transform of the radio path.
$S(f)$ represents the Fourier transform of the transmitted signal.
$N(f)$ is the Fourier transform of the noise.

The first part of the equation, represents the attenuation and phase shift on the path, while the second part represents the noise contribution of that path.

Beam-forming and MIMO both involve:

- gaining knowledge of this matrix

- Altering the emitted signal in such a way it has the desired characteristics for a targeted receiving station.

Many features in 802.11ac are aimed at simplifying the mechanisms of "beam forming" designed for 802.11n. Firstly, the interoperability of the "beam-forming" in the 802.11n standard is difficult between the different manufacturers that support each as a sub set of the MCS and other options.

In MIMO, estimation of the capacity of the radio path between two stations, is required for each pair of antennas, one for emitting, and the other downstream for receiving. The channel estimation corresponds to the determination of characteristics such as propagation delay, fading, crosstalk between paths, noise level, phase-shift compared to other channels. Actually we should talk about 'estimation of path' to avoid confusion with the concept of channel which is a subset of the 5 GHz band, but this confusion between the terms "path" and "channel" persists. In this document we will try to use the terminology ' radio path' whenever possible. This estimation of the radio path can be done with the LTF ("long training field") field. That's why the IEEE 802.11n amendment introduced specific fields for high throughput and MIMO (HT-LTF) and 802.11ac has introduced its own fields.

In 802.11ac this mechanism has evolved considerably. Some features are aimed at simplifying the mechanisms of "beam forming" of 802.11n. First of all, the "beam-forming" in 802.11n interoperability is difficult between the different manufacturers that support each as a different subset of the MCS and other options.

MIMO is possible if there is a knowledge of the radio path in the beam forming station before emitting, which makes it possible to twist the radio signals to attain the desired target. So a station forming a beam HT must have an accurate estimate of the path on which it is broadcast. Therefore a station forming a beam HT needs to calculate a FFT matrix appropriate for spatial processing during transmission toward a particular direction a beam HT training institution station, There are several ways to get this knowledge:

1. Because a beam-forming station has a knowledge of the parameters of the channel. We name this "implicit knowledge of the radio path". In this case the beam-forming station estimate the channel from past PHY frames received from the beam-formee station. Indeed this works only if the channel characteristics are the same in each direction. Often it's not the case, at least if no very complex processing is done.

2. Because it has a feedback indicating what the receiving station measure of the state of the channel. It is referred as explicit knowledge. In this case the beam-former station asks to the beam-formee station what the channel characteristics are. The beam-formee station measures the characteristics and sends them back in a report.

3. A close mechanism is "link adaptation" where some basic characteristics about the radio path are reported up-link.

Link adaptation is a related mechanism that can be supported by an immediate or delayed response as described below. An unsolicited MFB is also possible.
- Immediate: An immediate response occurs when the station receiving a MFB transmits the response in the TXOP obtained by the holder of the TXOP. This approach allows the applicant the MFB to obtain the benefit of the adaptation of the link within the same TXOP.
- Offline: a deferred response occurs when the station receiving a MFB transmitting response in the role of a TXOP owner in response to an MRQ in a previous TXOP obtained by the applicant of MFB.

Spontaneous: An unsolicited response occurs when the station sends a MFB independent of any previous MRQ.

Implementation of MIMO and beam-forming involves computation of the steering matrix (transmitter weights applied to the transmitted signal) used to steer the signal for a specific client. The weights, in turn, are derived from knowledge of channel, also known as Channel State Information (CSI). Actual implementation varies from chip vendor to chip vendor.

4.6.3.2 Beam forming Feedback

The beam-former and beam-formee stations can work together to educate each other on the characteristics of the MIMO channel. VHT-LTF symbols can be used to measure the channel for the space-time streams intended for the beam receiving station and can also be used to measure the channel for the interfering space-time streams. To successfully demodulate the space-time streams intended for the beam receiving station, the beam emitting station may use the channel state information for all space-time streams to reduce the effect of interfering space-time streams.

4.6.3.3 Beam forming Implicit feedback

In general, beam forming does not require the station receiving the beam to be aware of beam forming occurring at the station emitting the beam.

In this case, the station receiving the beam cannot and does not provide any feedback to help improve directivity. Assuming the channel is identical in uplink and downlink directions (i.e. reciprocal), the station emitting the beam then estimates the channel at its end based on receive signals. It uses these estimates to generate the steering matrix used to steer the outgoing signals. the primary advantage of this method is that beam forming can be done regardless of the station receiving the beam being an 802.11ac or a legacy device, also, it does not add a feedback overhead in the system. Thus, the overall gain achieved can be significant. The station emitting the beam transmits a training request (TRQ), which is a standard packet in 802.11, and expects to receive a sounding packet in response. Upon receiving the sounding packet, the station emitting the beam estimates the receive channel and computes the steering matrix that it will use to steer subsequent transmissions to this station receiving the beam in the transmit direction.

This method, however, requires computation of correction matrices in order to eliminate any mismatch between the uplink and the downlink channels; in other words, it requires calibration to maintain the reciprocity of the channel.

4.6.3.4 Beam forming and Calibration
Differences in the amplitude and phase characteristics of the transmit and receive chains associated with individual antennas degrade the reciprocity of the channel and cause performance degradation of implicit beam forming techniques. If the channel is reciprocal, the beam-former can use the training symbols that it receives from the beam-formee to make a channel estimate suitable for computing the transmit steering matrix. Generally, reciprocity requires calibrated radios in MIMO systems, hence the need for a calibration procedure.

4.6.3.5 Beam forming Explicit feedback:
In explicit beam forming, the station receiving the beam estimates the channel upon transmission of a sounding packet by the station emitting the beam. Depending on the implementation, the station receiving the beam responds with raw channel estimates or computes the steering matrix and feeds it back to the station emitting the beam in compressed or uncompressed forms. In case raw channel estimates are transmitted, the computation of the steering matrix occurs at the station emitting the beam.

This method provides very reliable steering matrices since the actual channel between the station emitting the beam and the station receiving the beam is estimated.

4.6.3.6 Beam forming and PHY preamble.
A different HT-LTF is transmitted by each spatial stream, to learn the each CSI (channel state information). If a station transmits identical symbols on different paths, the crosstalk

between paths can induce problems. For example if identical HT-LTF are passed through all antennas, a correlation can be detected leading to the false conclusion that there are a lot of cross-talk between paths. To prevent the this a different coding (called "cyclic shift diversity" CSD) is applied to symbols on the various antennas. As a result, the automatic gain control must be done by analyzing the HT-STF field before analyzing the HT-LTF field.

The receiver needs to know how many HT-LTF have been sent. This information is contained in the SIG field ("Signaling field") for high-throughput (HT-SIG) from the part of PHY preamble (with MCS, length, GI, and encoding information). The integrity of this signaling field is controlled by an 8-bit CRC and offers much greater protection than the single L-SIG parity bit.

The HT-LTF is pre-coded when using beam forming. The resulting stream can thus be estimated and used to decode the pre-coded data symbols.

In a DL MU-MIMO transmission, the LTF are considered resolvable when the access point transmits enough LTF to a station to estimate the channel of all space of each recipient. To enable an interference cancellation in a station during a transmission DL MU-MIMO, an access point can transmit the preamble using resolvable LTF.

4.6.4 SOUNDING PROTOCOL
4.6.4.1 802.11ac Sounding PPDU

To fully exploit the variation of the MIMO channel and transmit a beam on a MIMO link, a station may ask to another station to report about the characteristics of its MIMO channel and a send a MFB.
A station forming a beam uses the information in the response that it receives from a station receiving a beam forming request, to calculate a matrix of beam forming for the transmitter.

Supported formats are advertised in the "HT Capabilities" element of a station receiving a beam forming request.

Null Data Packet (NDP) is indicated by zero in the Length field in the HT-SIG and with the "Not Sounding" field set to 0.

A sounding PPDU is identified by setting the "Not Sounding" field in the High Throughput Signal field of the PHY header (HT-SIG) to zero.

4.6.4.2 Sounding protocol for mono-user

A sounding protocol is required for MU-MIMO and beam forming, the station emitting a beam needs to acquire the channel (CSI) status information to derive the appropriate pre-coding matrix that could be used to optimize the reception on one or more stations receiving a beam. In the IEEE 802.11n standard, the multiplicity of options for sounding protocol lead to poor interoperability between different chipsets involved in beam forming. As a result, IEEE 802.11ac uses a unique sounding protocol based on data packets using no data (NDP) to modelize CSI on each sub-carrier and send back a report.

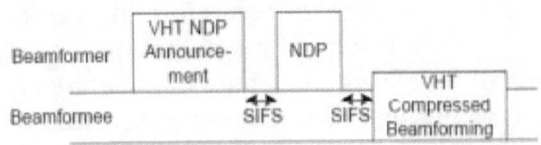

Figure 78: Single User sounding protocol

A NDP is a PPDU frame without PSDU, i.e. without data. This frame is generated in the PHY only layer, to meet the needs to estimate the radio path. The station emitting a beam announces it through a NDP announcement (NDPA) frame. Actually a station emitting a beam published the address of the stations that will receive a beam. The designated stations, therefore prepare to receive the upcoming NDP frame, and therefore calculate its beam-forming matrix using the singular value decomposition. The frame exchange is punctuated by the short inter-frame gap (SIFS). Upon received of the NDP, the station receiving a beam responds with a compressed version of the V. pre-coding matrix

Compression is used to reduce the sounding protocol overhead, while minimizing the loss of quantification. The beam-forming weights are therefore expressed in polar coordinates and quantified of the angles Ψ and Φ. The angular information is needed for each pair of antennas for transmitting-receiving and for each sub-carrier. The grouping of the sub-carriers can be used to further reduce protocol overload.

The duration of the sounding procedure depends on parameters such as the number of spatial streams, antennas, and bandwidth.

4.6.4.3 Sounding protocol for multi-user

The principle that the receiving stations (except the first) send back sounding reports is also used with the sounding protocol.

Indeed, it is thanks to the group ID contained in the NDPA, that stations receiving a request for beam forming can prepare itself for the upcoming NDP frame. The first station, as established by the ID of the group, follows the single user behavior, it sends its report as soon as possible, without being solicited by the sending station. The access point then queries the rest of the stations to get their beam-forming matrices through beam-forming sounding report frames.

With MU-MIMO, the matrix weights of the antennas are much more sensitive to changes in the channel. In the case of beam-forming single user, if the weights of antennas are outdated, system performance degrades to the case without beam forming. However, with MU-MIMO, if matrix antenna weights are not exactly what they should be, the emitting station introduces interferences on other stations, which leads to a degradation of the signal. This happens because the MAC mechanisms to manage MU-MIMO still expect the MU-MIMO to work perfectly. This implies that channel soundings should be done more often when using MU-MIMO that when the beam forming is done in single user mode. Therefore, in addition to the number of antennas and bandwidth, the number of stations receiving a request for beam forming (number of groups and the number of receiving stations per group) has a significant impact on the protocol overhead.

4.6.4.4 MU-MIMO Pre-coding

The use of MU-MIMO implies different pre-coding of each stream of each station. Schematically this pre-coding is to use to offset the shortcomings of the radio media and path length differences, so as to obtain a strengthening of the signal and only when and where it is desired. It therefore adjusts the amplitude, phase and delay of each spatial stream from the information collected during the sounding process. Pre-encoded channels must be estimated by each receiving station, because different matrices of beam forming are used. Thus VHT-LTF and the following fields are all pre-coded when MU-MIMO is used.

The access point must have a precise knowledge of each of the affected radio paths. Therefore, several soundings of channels are therefore made on a regular basis, to avoid protocol overloading. Another consequence is that groups of stations are to be defined by access point. The station group is based on the identifier allowing the receiving station stations to retrieve their data from the MU-MIMO received PPDU. Finally, the addressing procedure is also adapted to enable receiving station stations to recognize their PSDU. As a result, it can be clearly seen that MU-MIMO has implications both for PHY and MAC layers.

5 LINUX AND IEEE 802.11AC

5.1) INTRODUCTION

First the reader must be aware of the fact that there is not straight implementation on any 802.11 standard in Linux, nor any clear kernelland architecture despite an excellent web site (http://wireless.kernel.org).

Linux had sometimes been described as a huge bag of drivers, as the core of the kernel code is only a tiny part of the code. In an attempt to introduce some discipline, the interfaces between applications and drivers are done by dedicated layers, the applications calls some middleware and the hardware drivers register to the same middleware, resulting in applications been able to use hardware.

Linux had several interfaces for 802.11 drivers and applications. An obsolete one is "Wireless extensions" and the current one is cfg80211. In addition most 802.11 drivers do not implement their own MAC layer, instead they use the mac80211 layer, in this case they are called "SoftMac".

The AP functionality is implemented in the application layer, not the kernel. "hostapd" is a user space daemon for access point and authentication servers. It implements IEEE 802.11 access point management, IEEE 802.1X/WPA/WPA2/EAP Authenticators, RADIUS client, EAP server, and RADIUS authentication server.

Applications that use "Wireless extensions" can't use cfg80211 and mac80211 and vice-versa. "Wireless extensions" won't be studied in this book. This book will briefly describe the cfg80211 which is the API for applications that want to use or manage 802.11, it will describe also mac80211 which is the layer implementing 802.11 MAC.

Another words of caution to the reader is that neither mac802.11 nor the 802.11 drivers respect, even roughly, the 802.11 specification structure and concepts. They are not even well commented, probably instead to implementing the specification, the code evolved from a small initial kernel and features were added incrementally without much thinking as long it worked, an example being references to the 2001 area 802.11d. Other factors may be that companies even when they open source their drivers, were a bit reluctant to give the rationals behind the code. Independent coders have to reverse engineer the binaries, so they don't have access to those rationals. Anyway lot of Wi-Fi hardware uses Linux, so it seems to not be so bad after all.

5.1.1) INTERFACES

We are not interested in the inner workings of any specific 802.11ac hardware, because it is never generic. Even products from the same manufacturer are incompatible at the low level programmatic interface.

However a chip needs to interact with the Linux operating system of the host computer, via a driver and must interact with other 802.11 chips. So if we want to create a 802.11ac driver or even create a new 802.11ac hardware we must implement the following interfaces:

(1) Interface with the host

* configuration and initialization is done via cfg80211 Linux interface

* MLME (management) is done via the file mlme.c

* Interface with the LLC/MAC host layer. Here, as it is assumed that we use Linux and that the driver is Open Source, it means it is a SoftMac driver: the MAC layer is implemented by Linux's mac80211. What interests us is therefore the interface MAC with driver

(2) Interface with other 802.11 cards

(2-1) at the MAC layer:

* the question does not arise, except in terms of dot11... (Which tells if 802.11 functions are present or absent)? Linux do not implement an explicit 802.11 MIB so the what is permitted by the hardware is a bit tricky.

(2-2) Interface at the PHY level

* VHT preambles must be decoded

* the NDP must be processed

* Same for explicit beam-forming report

5.1.2) 802.11AC LINUX DRIVER SOURCE CODE.

Most of 802.11 Linux drivers are not 100% open source, in fact only a few 802.11n Atheros card have a Linux driver which is open source. Most of the time there is an upper layer which is open source, but the bulk of the code is implemented in a binary layer. Some 802.11 hardware are even accessed through a serial modem interface, removing any need for 802.11 drivers.

Anyway it was out of question to publish the full source code of a 802.11ac Linux driver as it would have required thousands pages of hard to understand code.

5-2) CFG80211

cfg80211 is the configuration API for 802.11 devices in Linux. It bridges userspace and drivers, and offers some utility functionality associated with 802.11. cfg80211 must, directly or indirectly via mac80211, be used by all modern wireless drivers in Linux, so that they offer a consistent API through nl80211. For backward compatibility, cfg80211 also offers wireless extensions to userspace, but hides them from drivers completely.

Additionally, cfg80211 contains code to help enforce regulatory spectrum use restrictions.

5-2-1) DEVICE REGISTRATION

In order for a driver to use cfg80211, it must register the hardware device with cfg80211.

The fundamental structure for each device is the '"wiphy structure"', of which each instance describes a physical wireless device connected to the system. Each such "wiphy structure" can have zero, one, or many virtual interfaces associated with it, which need to be identified as such by pointing the network interface's ieee80211_ptr pointer to a structure wireless_dev which further describes the wireless part of the interface.

Each "wiphy structure" structure contains device capability information, and also has a pointer to the various operations the driver offers. Its field members are described below:

5.2.2) SCANNING AND BSS LIST HANDLING

The scanning process itself is fairly simple, but cfg80211 offers quite a bit of helper functionality. To start a scan, the scan operation will be invoked with a scan definition. This scan definition contains the channels to scan, and the SSIDs to send probe requests for (including the wildcard, if desired). A passive scan is indicated by having no SSIDs to probe. Additionally, a scan request may contain extra information elements that should be added to the probe request. The IEs are guaranteed to be well-formed, and will not exceed the maximum length the driver advertised in the "wiphy structure" structure.

When scanning finds a BSS, cfg80211 needs to be notified of that, because it is responsible for maintaining the BSS list; the driver should not maintain a list itself. For this notification, various functions exist.

Since drivers do not maintain a BSS list, there are also a number of functions to search for a BSS and obtain information about it from the BSS structure cfg80211 maintains. The BSS list is also made available to userspace.

5.3) MAC80211

mac80211 is a framework which driver developers can use to write drivers for SoftMAC wireless devices.

SoftMAC devices allow for a finer control of the hardware, allowing for 802.11 frame management to be done in software for them, for both parsing and generation of 802.11 wireless frames. Most 802.11 devices today tend to be of this type, FullMAC devices have become scarce.

mac80211 implements the cfg80211 callbacks for SoftMAC devices, mac80211 then depends on cfg80211 for both registration to the networking subsystem and for configuration. Configuration is handled by cfg80211 both through nl80211 and wireless extensions.

In mac80211 the MLME is done in the kernel for station mode (STA) and in userspace for AP mode (hostapd).

If you have new userspace utilities which support nl80211 you do not need wireless-extensions to support a mac80211 device.

www.ingramcontent.com/pod-product-compliance
Lightning Source LLC
Chambersburg PA
CBHW020729180526
45163CB00001B/172